Site safety

SITE PRACTICE SERIES

General editors: Harold Lansdell, FCIOB, FCIArb, and Win Lansdell, BA

Site safety – *Jim Laney*
Site engineering – *Roy Murphy*

Books to be published in the Series
Making and placing concrete – *Edwin Martin Baker*
Timber frame housing – *Jim Burchall*
Security on site – *Leonard Earnshaw*
Site carpentry and joinery – *Keith Farmer*
Site relations – *Tom Gallagher*
Careers in the building industry – *Chris and Lynne March*
Fixings, fastenings and adhesives – *Paul Marsh*
Glazing – *Stanley Thompson*
Steel reinforcement – *Tony Trevorrow*
Exercises in brickwork and blockwork – *Arthur Webster*

Site safety

J.C. LANEY

Construction Press
LONDON AND NEW YORK

Construction Press
Longman House
Burnt Mill, Harlow, Essex, UK

A division of Longman Group Limited

*Published in the United States of America
by Longman Inc., New York*

First published 1982

British Library Cataloguing in Publication Data
Laney, J.C.
Site safety. – (Site practice series)
1. Building sites – Safety measures
I. Title II. Series
624 TH375
ISBN 0-582-40601-3

Library of Congress Cataloging in Publication Data
Laney, J.C., 1930–
Site safety.

([Site practice series])
Includes index.
1. Building sites – Safety measures. I. Title.
II. Series.
TH443.L23 690'.22 81-19540
ISBN 0-592-40601-3 AACR2

Printed in Hong Kong by
Astros Printing Ltd

Contents

Foreword

The Health and Safety at Work etc. Act 1974 and its subsequent legislation has been aimed at encouraging industry – both employers and employed – to adopt a 'self-regulatory' attitude to all matters relating to the well-being of the workforce and effective methods of work. This approach is more concerned with properly identifying potential dangers in the type of work carried out, rather than with merely satisfying detailed minimum statutory requirements.

This more positive approach to health and safety requires a correspondingly determined effort to ensure that everybody involved in the work process, at whatever level, is provided with essential information and guidance relative to the task being undertaken.

Numerous publications have been produced to this end, each with its own specific objectives and target audiences. Jim Laney's book *Site Safety* appears designed most usefully to fill an evident gap in the scope of existing publications. As stated in the first chapter of the book 'The key figure in site safety is the supervisor' and this is a level of site control which up until now has not been well covered.

Written in an easy style, the book is comprehensive without being too wordy, and contains the essential background information the supervisor needs to be effective and to fulfil his own responsibilities.

I commend this book to supervisors and to those responsible for instructing or advising them. It is construction safety guidance which will justifiably complement the existing range of advisory literature for the industry.

John M Lomax
NFBTE Director – Safety and Health

Acknowledgements

We are grateful to the following for permission to reproduce copyright material:

Figures 3.1, 3.2 and 15.4 reproduced by permission of the Controller of Her Majesty's Stationery Office; Figure 6.3 from British Standard CP 6031: 1981 and Figure 8.2 and 8.4 from CP 3010: 1972 reproduced by permission of the British Standards Institution, 2 Park Street, London, W1A 2BS from whom complete copies can be obtained; Figures 8.1, 8.3, 8.5 and 8.6 from *Construction Safety Manual*, published by BAS Management Services for the National Federation of Building Trades Employers, 82 New Cavendish Street, London, W1M 8AD; Figures 18.2 and 18.3 from Guidance Note GS.6 issued by the Health and Safety Executive, Baynards House, 1 Chepstow Place, Westbourne, London, reproduced by permission of the Controller of Her Majesty's Stationery Office; Figures 21.1, 21.2 and Table 21.1 from *Explosives, the Sale, Storage and Conveyance by Road*, reproduced by permission of Nobel's Explosives Company Limited (ICI); Figures 22.1 and 22.2 from *Code of Practice for Site Radiography* published by Kluwer Publishing Ltd., copyright the Oil and Chemical Plant Constructors Association; Figures 27.1 and 27.2 from *A Matter of Life and Death* published by Messrs. Skilton and Shaw, 115 Old Street, London EC1; Figures 28.1 and 28.2 from *Fire Safety Data NB1 – The Physics and Chemistry of Fire* published by the Fire Protection Association.

Part A

The safety scene

1

Safety and the supervisor

The key figure in site safety is the supervisor. He is the link between top management and the operatives and has responsibility for the development of proper safety attitudes and the detection and correction of unsafe working conditions and practices. To be effective in these duties he must be adequately trained in the techniques of accident prevention, have an appreciation of the many legal requirements now affecting site operations and understand some of the hazards that can arise.

This short book aims to provide some of this necessary knowledge and, as it were, point the reader in the right direction.

What is a 'safe' site?

Many definitions of 'safety' have been offered over the years but I prefer that which suggests that a construction site is safe when persons can go about their normal daily work without undue risk. This, I feel, is a realistic definition, for it accepts that there are risk situations in all everyday activities and does not pretend that a workplace can be entirely accident proof.

What then is an accident?

Probably the simplest definition of an accident is 'an uncontrollable occurrence which results in injury or damage'. The events leading up to an accident are, of course, controllable in most cases and this is what 'safety' is all about. The controlling of work situations by providing safe conditions and insisting on the use of safe working methods and procedures is the art of accident prevention.

Causes of accidents

A study of the causes of accidents is a necessary prelude to attempting to reduce the numbers that occur but one must dig deeper than the usual accident analysis, which lists causes such as falls of persons, objects falling, handling, etc.

Such lists are useful for statistical comparisons of like with like but do nothing to prevent accidents. In my view most 'accidents' arise from one of the following root causes:

Carelessness or couldn't-care-less attitudes

This is especially true of young workers. Fresh from school, college or university and full of enthusiasm, they often tear into what they are doing with little thought.

Ignorance and lack of training

This is probably the biggest single cause. In my experience too many assumptions of knowledge and ability of operatives are made by supervisors and by the same token too many operatives will happily tackle jobs in which they are inexperienced and untrained, with the inevitable result.

Lack of discipline

No one on a construction site expects military style discipline, but there is a need for a proper respect for authority. The supervisor must be prepared to crack down on wrongdoers and also to set the correct example. This is particularly important when horseplay occurs on site, as many accidents can be related to horseplay that has gone too far.

Distraction

A wood machinist, distracted by a call from a workmate, fed his hand instead of a piece of timber into the spindle moulding machine. The man had years of experience, was fully aware of the dangers, was well disciplined and reasonably safety conscious but he just hadn't reckoned on the effect of a moment's distraction. So in this situation there appears to be a dual need. We need to train our workers in job concentration and also to teach them not to distract their fellow workers.

Poor communications

By no means the least of the causes of accidents is poor communications. A badly given instruction or a glossed-over explanation can lead to mistakes which are costly in time and materials and which may lead to accidents. The foreman, holding a stake, who instructs a labourer 'When I nod my head, you hit it' gets what he deserves.

The golden rules for communicating your message are:

1. Be specific and not vague.
2. Use simple short words.
3. If possible illustrate the point in some way. The more senses that can be brought into use then the better chance there is of understanding being achieved.
4. Check that your instructions have been understood. Communication is a two-way process – an instruction has not been satisfactorily given until it is understood. An old RSM once said: 'First I tells 'em what I'm going to tell 'em, then I tells 'em, then I tells 'em what I've told 'em'.

Attention to these four commonsense rules will certainly improve communications on your site.

I would suggest, therefore, that these root causes – carelessness, ignorance, indiscipline, distraction and poor communications are behind most site accidents and that the major responsibility for bringing about an improvement falls, as do a good many responsibilities, into the lap of the supervisor.

Cost of accidents

The prime reason for the prevention of accidents must be a concern for the health and well-being of our work-force. No one wants to be injured or to see a fellow worker injured or, probably even worse, to feel responsible for someone's injury.

That being so, it is still important to appreciate fully what accidents can cost in terms of cash. The analysis of a simple injury accident will surprise many. Take a serious look at a recent accident that happened on your site; consider the headings that follow and roughly estimate the cost. Generally accident costs can be broken down into:

1. Wages paid to, but not earned by:
   (a) the injured man;
   (b) other operatives who stop work from curiosity, from sympathy or to assist;

(c) general foreman, trades foreman, timekeeper, etc., in helping, investigating, reporting the accident and arranging for continuity of work.

2. Taking on and instructing replacement labour.
3. Interruption of the planned sequence of work.
4. Damage to equipment, plant and materials.
5. Plant and transport standing out of use.
6. Administrative expenses – first aid, phone calls etc.
7. Initial lower productivity when injured man returns to work.

Add to these the administrative costs of your head office staff-safety, insurance, etc., and you may be surprised.

Accident investigation

The purpose of accident investigation is to establish clearly *what* caused the accident so that action may be taken to prevent a recurrence.

Accident investigation is *not* immediately concerned with apportioning blame. The person carrying out the investigation must always be tactful, patient and unbiased and should undertake the investigation as soon as possible after the accident occurs. Facts become blurred after a time and imagination often plays a larger part than memory. In an accident investigation certain basic facts are required to be established about the injured person:

1. Name, sex and age.
2. Address, marital status and occupation.
3. Time and date of accident and time and date of ceasing to work as a result of the accident (not always the same).
4. Nature of injuries sustained and/or details of damage to plant or equipment.

Having obtained these basic facts the person carrying out the investigation should then proceed to establish other aspects in more detail in an attempt to identify the 'true' cause. He should establish:

1. The exact location of the accident.
2. The exact operation that the injured person was engaged on.
3. Whether this was in accordance with the instructions he had been given.
4. Exactly what instructions had been given and by whom.
5. Whether the injured person had carried out this particular

operation previously and what training/job instruction he had received.

6. If special safety equipment was necessary for the operation, had it been provided and if provided had it been used.
7. The sequence of events leading up to the accident:
   (a) was some material or equipment being handled, if so establish size and weight;
   (b) was safe access provided – if the person injured fell then establish the distance of fall;
   (c) was plant involved and if so was it properly guarded and safe for use.
8. Whether the injured person was in any way disabled prior to the accident
9. What was seen by witnesses.
10. What, if any, statutory regulations had been disregarded and whether the injured person was aware of this fact.
11. If any other contractors' employers' men, materials or plant were involved.

Although this list may not be complete, the answers obtained should enable the person carrying out the investigation to arrive at a reasonably accurate picture of what actually occurred. Any information so gained should be put to good use in providing for such action as is necessary to prevent a recurrence.

Prevention of accidents

Accidents happen to, and are caused by, people or their faults and much can be achieved in preventing accidents by bringing about a change of attitude on the part of the work-force and by giving them a better understanding of the root causes. The supervisor has a major role in this process. He must combat more effectively the careless and couldn't-care-less approach to the job; he must not necessarily assume knowledge on the part of the operative; he must give particular attention to the young and to workers new to the industry; he must be prepared to tackle indiscipline; he must improve his communications technique and he must acquire a satisfactory technical knowledge of both statutory regulations and safe working methods and techniques.

Armed with this knowledge, and with a determination to 'manage' his site, he can make a real contribution to safety.

2

Legal requirements

The supervisor on site is the representative of the employer and as such he needs to be aware of the main 'employer' responsibilities, both in common law and in statute law.

Common law and statute law

Common law is the ancient law of England and is not to be found written in any one document but in the reported decisions of judges. Over the years, decisions made by the high court have modified and extended the common law but, as it relates to the workplace, it is still basically a set of ground rules describing a proper relationship between employers and employees.

Common law requires that the employer shall provide:

1. A safe place of work.
2. A safe system of work.
3. Safe access to and egress from the place of work.
4. Safe plant and equipment.

Statute law, which consists of Acts and Regulations introduced and approved by Parliament, is complementary to common law and is more precise in its demands, and any breaches of its requirements can lead to criminal proceedings before a magistrate's court, or in serious cases before a higher court.

If an employee sustains injuries he may seek compensation against his employer for loss of earnings and for pain and suffering. Claims such as these are made under common law, are generally known as 'civil' claims and are not directly associated with any action taken against an employer under statute law.

What the supervisor *must* appreciate is that failing to comply with statute law is a *criminal* offence and that in certain circumstances the supervisor may be personally charged with an offence in addition to the company.

The Health and Safety at Work etc. Act 1974

This Act has been described as the most important reform of Britain's industrial health and safety laws since Shaftesbury's Factory Act of 1833. It not only radically alters the scope and coverage of existing provisions and the way in which they are administered and enforced, but also aims to simplify the law by putting the onus upon self-regulation within industry itself. The principal idea is to get away from the concept of negative regulation and, instead, the Act imposes a series of basic obligations upon employers and employees so as to place the primary responsibility for doing something about occupational hazards upon those people who actually create the risks and those who work with them.

The Act is split into four parts:

I. Health and Safety at Work Provisions;
II. Employment Medical Advisory Service;
III. Building Regulations;
IV. Miscellaneous and General Provisions.

Part I will be of most interest to us and the following notes are limited to the most important sections therein:

Section 1 sets out the general purposes of Part I of the Act as:

1. Maintaining or improving standards of health, safety and welfare at work.
2. Protecting other people against risks to health and safety arising out of work activities.
3. Controlling the storage and use of dangerous substances.
4. Controlling certain emissions into the air from certain premises.

Section 2 places a general duty on employers to ensure the health, safety and welfare at work of their employees; to provide them with information, instruction and training; to consult them concerning arrangements for joint action on health and safety matters; and, in prescribed circumstances, at the request of duly appointed safety representatives, to establish safety committees; to prepare and publicize a written statement of their safety policy and arrangements and to include safety performance reports in company annual reports.

Section 3 places a duty on employers and self-employed to ensure that their activities do not endanger *anybody* (general public) and, in prescribed circumstances, to provide information to the public about any potential hazards to health and safety.

Section 4 places a duty on anybody responsible for places of work to ensure that the premises themselves, as well as plant and machinery in them, do not endanger people using them.

Section 5 places a duty on anybody responsible for places of work to prevent dangerous emissions.
(*Note*: The emissions are further described as noxious or offensive and could include noise).

Section 6 places duties on designers, manufacturers, importers and suppliers to ensure that articles and substances are safe if used in accordance with information supplied. (This also applies to installers of plant.)

Section 7 places duties on employees to take reasonable care that they do not endanger themselves or anyone else who may be affected by their work activities.

Section 8 places a duty on all persons not to misuse anything provided in the interests of health and safety purposes.

Section 15 is the section of the Act which empowers the Secretary of State to make Health and Safety Regulations for any of the general purposes of the Act. The Secretary of State's powers to make Statutory Orders are defined in Schedule 3 of the Act and are most extensive.

Section 16 deals with the production of 'codes of practice'.

Section 17 describes the status which approved codes of practice will have in legal proceedings and, in practical terms, there appears to be little difference in status between codes of practice and regulations proper.

However, it is expected that the number of actual court proceedings will diminish under the new Act. Although the threat of criminal proceedings is retained, as it were, as a last resort, the main sanction under the new Act will be administrative. I refer to 'Improvement Notices' and 'Prohibition Notices'.

Section 21 empowers an Inspector to issue an improvement notice where a person contravenes a requirement, requiring him to remedy the contravention within a specified period. Failure to make the necessary improvements constitutes an offence.

Section 22 empowers an Inspector to issue a prohibition notice where, in his opinion, an activity gives rise to risk of serious injury. Prohibition notices may specify remedial measures to be taken and may either fix a time after which the activity will be prohibited unless such measures are taken, or may require the activity to cease immediately until the cause of the risk is remedied.
(*Note*: Inspectors are required to advise employee representatives of the content of improvement and prohibition notices).

Section 24 provides for appeals against improvement and prohibition notices to be made to an industrial tribunal. An appeal against an improvement notice will have the effect of placing the notice in abeyance pending determination of the appeal.

An appeal against a prohibition notice, however, will not have this effect and the notice will continue in force pending determination.

Section 33 lists offences created by the Act and the type of penalty which may be imposed.

Section 36 says that if an offence is committed by a person but it can be proved that the offence was due to the fault of someone else, then that other person may be charged and convicted of the offence whether or not the original person is prosecuted.

Section 37 makes provision for the prosecution of a director, manager, secretary or *similar officer* of a company, as well as the company, if it can be proved that the offence was with his consent or knowledge.

The sections quoted above are probably the most important of which the supervisor should be aware but further study of the Act in more detail is to be recommended. There are two specific points of Section 2 which need some elaboration.

Safety policy

The statutory instrument introduced by the Secretary of State in respect of this part of section 2 requires all employers, except those having less than five employees, to prepare a written statement of their *Safety Policy* and the organization and arrangements for carrying out that policy.

It is to be hoped that all supervisors reading this book will have received a copy of the policy of their employer; the important point is to appreciate fully the significance of this policy document and the possible effect of not complying with its requirements.

Most policies for construction companies will make a general statement of intent and then specify, in some detail, responsibilities for different levels of management. A fairly general set of responsibilities for a site agent or supervisor would be to:

1. Organize sites so that work is carried out to the required standard with minimum risk to men, equipment and materials – on larger sites issuing work method instructions in written form.
2. Know the broad requirements of the Construction Regulations and other relevant legislation.
3. See that the Construction Regulations and other legal requirements are observed on site; that all registers, records and reports are in order and that the 'competent person' appointed has sufficient knowledge of plant or machinery to evaluate all aspects of its safe operation.
4. Give all trades foremen and gangers precise instructions on their responsibilities for correct working methods; see that they do not require or permit men (particularly apprentices) to take unnecessary risks.
5. Arrange delivery and stacking of materials to avoid doubling risks by double handling; position plant effectively; ensure that the electricity supply is installed and maintained without endangering men and equipment.
6. Plan and maintain a tidy site.
7. Implement arrangements with subcontractors and other contractors on site to avoid any confusion about areas of responsibility.

8. Check that all machinery and plant, including power and hand tools, is maintained in good condition.
9. Make sure that suitable protective clothing is available where appropriate and that it is used.
10. Ensure that a qualified first-aider and all items of first-aid equipment as required by Construction Regulations are available and their location known to employees.
11. See that proper care is taken of casualties and know where to obtain medical help and ambulance service in the event of a serious injury. (Nominate others to act in emergency).
12. Accompany HM Factory Inspector on site visits and act on his recommendations.
13. Release supervisors and operatives, where necessary, for on- or off-site safety training.
14. Co-operate with the company safety officer; act on his recommendations.
15. Liaise with the Fire Brigade on fire prevention.
16. Set a personal example.

In the event of a serious accident on site the safety policy will, in many cases, be the starting point from which an inspector will embark upon his investigation. The supervisor must make sure that he both understands and carries out his responsibilities.

The safety policy statement will also inevitably be subjected to close scrutiny by workforce representatives which leads me to my second elaboration of part of Section 2.

Safety representatives

Following the implementation of this part of Section 2 the employers in the construction industry agreed with the unions, those who are party to the National Joint Council for the Building Industry (NJCBI) or to the Civil Engineering Construction Conciliation Board (CECCB) that a new working rule be devised in order to form an agreed basis for the application of the new requirements in the construction industry.

The rule is number 7A in the NJCBI agreement and number 18A in the CECCB agreement, and the main points are as follows:

Appointment

When a union proposes to appoint a safety representative at a place of work, the full-time official of the union concerned will normally discuss with the employer the number of safety representatives that may be appropriate for that workplace, and define the group or groups of operatives who are to be so represented.

A 'union' in this particular paragraph means, in building, a union which is party to the NJCBI, e.g. UCATT, TGWU, GMWU or FTAT; or, in civil engineering, a union which is party to the CECCB, e.g. TGWU, UCATT or GMWU. The number of safety representatives appropriate in any given circumstances is a matter to be determined locally and will depend on such factors as the extent of the workplace, the groups of operatives to be covered, the number of unions involved and any special features, such as the operation of shift systems. Discussions should always be with a full-time official of the union concerned. Where more than one NJCBI/CECCB union is involved, the employer should aim at combined discussions and agreement among the unions on the appointment of one safety representative at the workplace.

A safety representative appointed by the union shall be issued with appropriate credentials and the union shall notify the employer in writing, as early as possible, of the appointment made, indicating the groups of operatives to be represented. So far as is reasonably practicable, the safety representative so appointed shall be in the employment of the main contractor.

When proposing to appoint an operative as a safety representative, the trade union should take into account the requirements of the regulations relating to the experience and length of service of a person so appointed. The experience and length of service appropriate for appointment of safety representatives in the construction industry will vary according to the circumstances prevailing at the place of work concerned. The union shall normally discuss with the employer what is appropriate in individual cases. Twelve months' experience in the industry should normally be the minimum.

Functions

A summary of the functions of a safety representative is:

1. To investigate potential hazards and dangerous occurrences at the workplace and to examine the causes of accidents at the workplace.
2. To investigate complaints by any operative he represents relating to that operative's health, safety or welfare at work.

3. To make representations, which should normally be confirmed in writing, to the employer on
   (a) matters arising out of sub-paragraphs (1) and (2) above;
   (b) general matters affecting the health, safety or welfare at work of the operatives at the workplace.
4. To carry out inspections in accordance with 'Workplace Inspections' below.
5. To represent operatives at the workplace with Inspectors of the Health and Safety Executive and of any other enforcing authority.
6. To receive information from inspectors in accordance with Section 28 (8) of the Health and Safety at Work Act 1974.
7. To encourage co-operation between the operatives he represents and the employer on safe practices.
8. To attend meetings of safety committees, where he attends in his capacity as a safety representative in connection with any of the above functions.

Workplace inspections

The regulations make provision for a safety representative to inspect the workplace or a part of it to which his appointment refers, if prior notice, in writing, has been given to the employer of the intention to do so. These inspections will take place in accordance with a programme agreed with the employer, will normally be at intervals of not less than three months but may be more frequent if there have been substantial changes in the conditions of work or if new information has been published by the Health and Safety Commission. Facilities will be provided by the employer in accordance with the regulations and there are advantages in formal inspections being carried out jointly.

Inspection of documents

The regulations require employers to make available to safety representatives any information within their knowledge which is necessary to enable them properly to fulfil their functions and to inspect relevant documents except medical records of individuals. Entitlement to such information is subject to the exceptions set out in the regulations. Should any dispute or difference arise concerning the provision of information under this rule, the matter is to be referred to the joint machinery of the industry for settling disputes.

Safety committees

Where an employer is requested, in writing, to establish a safety committee by at least two safety representatives, he shall consult the union(s) party to the NJCBI/CECCB who have members at the workplace with a view to deciding what arrangements are appropriate, having regard to the nature of the workplace and the consultative arrangements which already exist. Consultation should cover, as appropriate, the membership of the committee (taking account of sub-contractors); the functions, procedure, and meeting programme of the committee; its interrelation with the safety representatives; the date of establishment; and the posting by the employer of a notice stating the composition of the committee and workplace(s) to be covered by it.

It is recognized that the structure and constitution of safety committees will vary according to local conditions. Therefore, where a request to establish a committee is made, the form that the committee will take must be subject to consultation in the context of the particular workplace concerned. Consultation is to involve all the NJCBI/CECCB unions who have members at the workplace, not just the safety representatives who initiate the request.

Training and time off

National Joint Council for the Building Industry and the Civil Engineering Construction Conciliation Board have approved a form of basic training, taking into account the functions of safety representatives placed on them by the regulations and the understanding between the CBI and TUC.

If, at the time of appointment, the safety representative has not already satisfactorily completed the training so approved, the employer will allow the safety representative paid time off to complete this training. The amount of paid time off shall be what is considered to be reasonable for the purpose, in accordance with arrangements to be agreed with the employer, having regard to the form of training approved by the NJCBI, CECCB and to the Code of Practice of the Health and Safety Commission.

The employer shall notify to the safety representative the management representative who is authorized to act for the employer under the regulations to whom he has access and shall, in consultation with the unions, arrange the procedure for such paid time off for the safety representative as may be necessary to carry out any functions previously described.

Other regulations

Whilst the Health and Safety at Work, etc. Act deals with general duties, the other codes of regulations that apply to construction work control specific operations. The four main codes affecting construction are:

The Construction (General Provisions) Regulations 1961;
The Construction (Lifting Operations) Regulations 1961;
The Construction (Working Places) Regulations 1966;
The Construction (Health and Welfare) Regulations 1966.

There are many other codes of regulations that are relevant to construction work as well as to industry generally; for example – Abrasive Wheels Regulations and the Protection of Eyes Regulations. Details of these and of the four construction codes quoted above will be given in the practical chapters that follow.

Powers of Health and Safety Executive Inspectors

Questions are often asked regarding the powers of HSE Inspectors, and supervisors should appreciate that these powers are considerable. HSE Inspectors may:

1. Enter 'premises', which include contract sites, and make examinations and investigations.
2. Take with them a police constable if obstruction is expected.
3. Direct that premises or any part thereof shall be left undisturbed to enable investigations to be carried out.
4. Take measurements, photographs, recordings and samples.
5. Require involved persons to answer questions and to sign a declaration of truth.
6. Require production of records, registers and documents.
7. Issue improvement and prohibition notices and take criminal proceedings against employers *and employees.*

Offences

Finally, supervisors should be aware of the offences for which they could be prosecuted; these are the main possibilities:

1. Failing to discharge a duty or contravening any part of Sections 2 to 9 of the Act.
2. Contravening any Health and Safety Regulation.
3. Preventing any other person from appearing before an inspector to answer his questions.
4. Contravening any requirement or prohibition imposed by a 'notice'.
5. Intentionally obstructing an inspector.
6. Intentionally making a false entry in a register or other document.

The penalties that can be imposed are not to be treated lightly and are:

1. On summary conviction in a magistrate's court – a fine not exceeding £1000 for each offence and,
2. On conviction on indictment – an unlimited fine or imprisonment for up to two years, or BOTH.

Better be safe than sorry!

Part B

Preparing for the contract

3

Planning for safety

No responsible supervisor will consider making a start on a particular operation on site until he has given some thought to sensible programming. To start mixing the mortar before the bricks have been ordered is a bit premature; planning, essential in all aspects of construction work, applies to safety matters more than most.

Legal notifications

The first important thing is to get off on the right foot with the Health and Safety Executive and to make sure that the appropriate notifications are made.

All contracts expected to last 6 weeks or more must be notified on Form F 10 within 7 days of starting work on site (see Fig. 3.1). Additionally, if the contract is expected to last for 6 months or more then a second form, an OSR1, must be despatched (see Fig. 3.2). It is usual for these administrative duties to be dealt with by the head office of your company but you are advised to check that this is the case. It is also recommended that you send a written notification to the local ambulance authority if you have more than twenty-five employees on site. No special form is provided for this purpose.

Statutory notices and registers

You should also make sure that you have the statutory notices ready for display and that all required registers and codes of regulations are available. The main notices that are required are:

1. Form F 3 – Abstract of Factories Act – building operations and works of engineering construction.

F 10
Reprinted December 1979

For official use
Registered...............................
Visited..................................

FACTORIES ACT 1961

Notice of building operations or works of engineering construction*

1	Name of person, firm, or company undertaking the operations or works.	
2	State whether main contractor or sub-contractor.	
3	Trade of the person, firm or company undertaking the operations or works.	
4	Address of registered office (in case of company) or of principal place of business (in other cases).	
5	Address to which communications should be sent (if different from above).	
6	Place where the operations or works are carried on.	
7	Name of Local Government District Council (in Scotland, County Council or Burgh Town Council) within whose district the operations or works are situated.	
8	Telephone No. (if any) of the site.	
9	How many workers are you likely to employ on the site?	
10	Approximate date of commencement.	
11	Probable duration of work.	
12	Is mechanical power being, or to be, used? If so, what is its nature (e.g. electric, steam, gas or oil)?	
13	Nature of operations or works carried on:	

(a) Building operations *(tick items which apply)*

Construction	of	Industrial building ....
Maintenance		Commercial or public building ....
Demolition		Dwellings over 3 storeys ....
		Dwellings of 3 storeys or less ....
		Others ....

(b) Works of engineering construction *(specify type)*

I hereby give notice that I am undertaking the building operations or works of engineering construction specified above.

Signature   Date

NOTE

* Any person undertaking any building operations or works of engineering construction to which the Act applies is required by the Act, not later than seven days after the beginning of any such operations or works, to serve on the Inspector for the district a written notice giving particulars specified in section 127(6) unless (a) they are operations or works which the person undertaking them has reasonable grounds for believing will be completed in less than six weeks, or (b) notice has already been given to the Inspector in respect of building operations or works of engineering construction already in progress at the same place. This form should be filled up and sent to HM Inspector of Factories for the district in which the operations or works are carried on.

572 8033375 200M 12/79 HGW 752

Fig. 3.1 Specimen Form 10 issued by the Health and Safety Executive, Baynards House, 1 Chepstow Place, Westbourne Grove, London W2 4TF.

OSR 1
Reprinted September 1977

OFFICES, SHOPS AND RAILWAY PREMISES ACT 1963

Section 49 of the Offices, Shops and Railway Premises Act 1963, and the Notification of Employment of Persons Order 1964 require that if on the 1st May, 1964, you are employing, or are intending after that date to begin to employ, persons to work in shop or office premises other than certain offices occupied by railway undertakings (for definitions see notes 2 to 6 on pages 2 and 4), you shall complete this form and send it to the appropriate authority (see note 1). Persons who are already so employing staff on the 1st May, 1964, should delete Part I, complete Part III and send it off before 31st July, 1964. Persons who are intending to begin so to employ staff after 1st May, 1964, should delete Part II, complete Parts I and III, and send off the form before they first begin to employ staff.

Please ensure that the duplicate form on page 3 is completed (using carbon paper if you wish) and sent with the top copy to the appropriate authority. They will send the duplicate form to the fire authority for the area who also have duties in connection with the Act. (You may need a fire certificate if more than a certain number of people are employed in your premises – see note 8.)

A separate form should be completed for each set of premises with a different postal address. Where several occupiers have premises at the same address, each occupier should complete a form in respect of his premises.

Further notes will be found on pages 2 and 4 describing the classes of office and shop premises within the scope of the Act and indicating the appropriate authority to whom this form should be sent. You are advised to read these notes before you complete the form.

NOTICE IN FORM PRESCRIBED BY THE SECRETARY OF STATE FOR EMPLOYMENT, OF EMPLOYMENT OF PERSONS IN OFFICE OR SHOP PREMISES

PART I

Notice is hereby given that on the .... (*insert date*), the employer specified in Part III of this notice, will begin to employ persons to work in the premises described therein.

PART II

Notice is hereby given that the employer specified in Part III of this notice is employing persons to work in the premises described therein.

PART III

1. (*a*) Name of the employer ....
   (*b*) Trading name, if any ....
2. (*a*) Postal address of the premises ....
   ....
   (*b*) Telephone no. ....
3. Nature of business ....
4. How many persons are or will be employed by the employer in office or shop premises at the above address in the following types of workplace? *(See notes 3-7.)*
   (*a*) Office ... ...
   (*b*) Shop (retail) ... ...
   (*c*) Wholesale department or warehouse... ... ...
   (*d*) Catering establishment open to the public ... ...
   (*e*) Staff canteen ... ...
   (*f*) Fuel storage depot ... ...
   TOTAL ... ...
   Of the TOTAL, how many are females? ... ...
5. How many of the total are or will be employed on floors *other* than the ground floor? ... ...
6. Of the total stated in reply to question 4, are any (or will any be) housed in separate buildings?... ... ... *(Answer Yes or No)* ....
7. Is the employer the owner of the building(s) (or part of the building(s)) containing the premises? ... ... *(Answer Yes or No)* ....
8. If not, state the name and address of the owner(s) or person(s) to whom rent is paid... ... ...

Signature of employer or person authorised to sign on his behalf ....Date

For official use

Fig. 3.2 Specimen Form OSR1 issued by the Health and Safety Executive, Baynards House, 1 Chepstow Place, Westbourne Grove, London W2 4TF.

2. Form F 954 – Electricity Regulations 1908 and 1944.
3. Form F 996 – Lead Paint Regulations.
4. Form F 2345 – Abrasive Wheels Regulations 1970.
5. Form F 2358 – Asbestos Regulations 1969.
6. Form F 2440 – Highly Flammable Liquids and Liquefied Petroleum Gases Regulations 1972.
7. Form F 2470 – Woodworking Machines Regulations 1974.
8. – Electric Shock Placard.
9. Form OSR 9 – Offices, Shops and Railway Premises Act 1963.

Items (1) and (9) are required on all construction contracts but the others are only technically necessary when work involves the equipment or material concerned.

Many companies maintain registers in company head offices and this is permitted in most cases. It is usual, however, to keep the following on site:

1. B.I. 510 – National Insurance accident book.
2. F 36 – General register for sites.
3. F 2509 – HSE accident register.
4. F 91 – Weekly inspection register.
5. F 2202 – Register of shared welfare facilities.

Site access and egress

Consideration must be given when planning the site to the positioning of access points and to the routing of site roads.

Site entrances should be, wherever possible, in a position where any inconvenience to the public or to road transport is kept to a minimum and suitable notices should be displayed warning other road users of site entrances. Particular thought should be given to the routing of traffic *on* the site. The provision of one-way systems and the avoidance of vehicles reversing is to be recommended on all sites, however small.

Stores compound

Storage areas and buildings need to be carefully sited and planned. Special regulations apply to the storage of Highly Flammable Liquids and Liquefied Petroleum Gas and Explosives and these items are dealt with in more detail in Chapters 13 and 21, but other stores items need careful consideration. Wherever possible timber should be stored not less than 6 m from any building to reduce the risk of fire spreading. The refuelling of plant from site storage tanks needs to be considered, adequate access maintained and fire fighting equipment provided. One of the main problems that can arise with badly organized stores areas is the curse of double handling. Stores incorrectly placed are blocked in by other materials and a chess-like operation is required to release them. Space is often short, tempers become frayed, risks are taken and injuries sustained. The storage of equipment that is in regular use should be given special consideration. Scaffolding components, props and struts, road forms, etc., all need special locations for easy and safe access.

Offices, canteens and welfare

Care must be taken to site offices, canteens and welfare buildings in such a way as to reduce fire risk to adjacent buildings and attention must be given to the means of escape. If it is to be a fairly large office with more than twenty employed, or more than ten employed above ground floor level, then a fire alarm system will need to be provided and a safe means of escape certificate obtained from the Fire Authority.

Heating and lighting will need to be considered and, if liquefied petroleum gas is to be used for this purpose, particular attention paid to positioning of supply bottles, installations and adequate ventilation. Heaters, of any type, must be protected to prevent clothing or materials being draped over or placed too close to them. This is particularly important in drying rooms.

Washing facilities and toilets must be planned for and provided at the start of the contract. Details of the legal requirements are given in Chapter 23 but these should be considered to be minimum requirements. Good welfare facilities are the first step to good site relations and subsequent good work performance.

Some years ago a senior Inspector of Factories, when dealing with this point in his lectures to site agents, used to say, 'When William landed at Hastings in 1066, in two days he had constructed a timber fortress for his 1000 men. It takes the average small builder three weeks to put up a mess hut for half a dozen men.'

I was never very sure of his historical accuracy but have many times been forced to agree with the second part of his analogy.

The provision of a suitable first-aid kit is vital and so is the need for someone who is trained in first-aid. Again the Health and Safety (First Aid) Regulations 1981 quote minimum requirements: 'Where a contractor has more than fifty men on site, the man in charge of the first-aid box must hold a valid first-aid certificate.' One accepts that there are degrees in everything and that to require a 'certificated' first-aider on *all* sites would be going to the extreme, but it must be sensible for the site supervisor to have at least a basic understanding of first-aid emergency treatments and priorities. This is another responsibility for the supervisor but one that is very worthwhile (see Ch. 27).

Plant

The plant and equipment to be used on site needs special consideration from the safety angle.

Before any 'lifting appliance' is used on site you must ensure that it has been examined as required by the particular regulations. The examinations vary depending on the type of plant but the most common are:

1. Excavators – thorough examination every 14 months.
2. Chains and lifting gear – thorough examination every 6 months.
3. Hoists – thorough examination every 6 months.
4. Cranes – thorough examination every 14 months; also a load test every 4 years or after substantial alteration or repair.

The only way the site supervisor can ensure that these examinations have been carried out is to require a copy of, or at least a sight of, the appropriate certificate. Some plant companies provide copies in the cabs of driver operated plant but not all do this. Make your requirement known to the plant company *before* the plant arrives and so avoid trouble and delay. For the same reason don't assume that necessarily all the ancillary equipment will automatically arrive on site with an item of plant. Remember to specify how many sets of gates you need for the hoist and don't leave it to the supplier to guess. If you do then the most you will receive will be two sets, one top and one bottom.

Services

There are two aspects to be considered in respect of services: first to establish what, if any, exist on the site and secondly to make provision for the site requirements.

Enquiries will have to be made of all the statutory undertakings for details of services which enter or cross the site area and arrangements agreed with these undertakings to protect or divert the services. The position of any underground services that are to be undisturbed should be clearly marked, particularly electricity cables. Adequate provision should also be made at the early planning stage for the protection of overhead cables. Plastic bunting on its own is not sufficient protection. This is a useful visual warning but in addition a physical barrier of greater strength should be provided. Wire rope of 12 mm diameter is useful for this purpose.

Services for site use must be planned

Electricity

Arrangements will need to be made well in advance for the supply of electricity, remembering that a transformer or transformers may be required when it is necessary to break down the supply to 110 V for supplying hand tools and temporary lighting.

Water

Water will be required as soon as the job begins, not only for the work itself but also for drinking, washing facilities and toilets. If it is possible to organize a temporary mains run in conjunction with the supply for the completed job this may be advantageous.

Telephone

Arrangements for summoning an ambulance should be displayed adjacent to the telephone so that no delay takes place. In remote areas where no telephone facilities are available there should be transport on the site capable of taking a stretcher case. Again I must say that this must be considered to be a minimum requirement.

Protective clothing and equipment

Associated with welfare but worthy of a special mention is the provision of protective clothing and equipment. Arrangements should be set up for ordering, storing, issuing, cleaning, recovery and supervision of use of:

1. Safety helmets.

2. Goggles.
3. Gloves.
4. Ear defenders.
5. Inclement weather clothing and rubber boots.

and in specialized work situations:

6. Safety harness, nets.
7. Respirators or breathing apparatus.
8. Life jackets, lifebuoys, grab lines and rescue boat.

Training

Finally the site supervisor must be reminded of his responsibilities for ensuring that his staff and operatives have received instruction and training in the jobs they will be required to undertake and, in addition to the general requirement of the Health and Safety at Work etc. Act, there is also a specific training requirement for any employees required to mount or change an abrasive wheel or cutting disc.

These specific requirements are covered in more detail in Chapter 10 but should always be considered with the other assessments of competency that are necessary at the planning stage of the contract.

4

Instructing for safety

One job the supervisor needs to do really well is to give instructions. If his instructions are good he can expect the work to be done well and safely but instructions that are bad, incomplete or unclear lead to bad workmanship and accidents.

Induction training

The initial training provided for the new entrant and especially the new young entrant is most important. Major companies will have training staff to perform this service but many smaller companies, and particularly those with small to medium sized contract sites, will leave this job to the supervisor.

Much can be achieved in even a short instruction period and the supervisor, however short of time he finds himself, should make every effort to arrange to familiarize new entrants with the common types of hazard on site and to describe the broad requirements of the company's safety policy to them before they commence work.

This should be considered to be an absolute minimum and a more detailed period of instruction should follow as soon as possible. The sort of headings that should be covered by the instruction are:

Hazards on site and simple precautions

(a) *Machinery*. Principles of starting and stopping plant, safe refuelling, guards on moving parts;
(b) *Transport*. Riding in unsafe positions, dangers of and from reversing vehicles, passing close to excavations;
(c) *Falls*. Proper use of access ladders, dangers of makeshift platforms, problems with weather – ice, mud;
(d) *Electricity*. Hazards of overhead and underground cables, use of transformers, dangers of makeshift connections;

(e) *Liquefied petroleum gas*. Basic properties, need for clips on hoses, explosion, fire and asphyxiation possibilities;
(f) *Fire*. Use of extinguishers, procedure to be taken if fire occurs;
(g) *Manual handling*. Basic principles, back straight, bent knees, good grip, etc;
(h) *Discipline*. Dangers of horseplay;
(i) *Housekeeping*. Need for tidiness, removal of nails from struck timber, brick stacks and stores areas.

Health and welfare

Describe site facilities, discuss potential health hazards – dermatitis, tetanus, leptospirosis, need for first-aid treatment.

Protective clothing and equipment

How to obtain particular items, helmets, gloves, goggles etc. What the policy is regarding wearing these items, legal requirement to wear goggles and to look after safety equipment.

Safety policy

Discuss company policy, responsibilities of supervisors and operatives, accident reporting procedures, need for reporting dangerous occurrences and the general role that the new entrant can play in helping create and maintain a safe working environment.

Job instruction

Each person starting a job should be instructed to do it properly. Even if he has, or says he has, performed similar work before, it is dangerous to assume knowledge.

The safety-conscious supervisor wants to make sure that the new employee will work safely and this might not have been emphasized where the person worked previously.

Instructing, like any other skill, can be learned by practice and it will be necessary for the supervisor to acquire this skill for himself and then to pass it on to his junior supervisors, trades foremen and gangers as they will be generally involved with the day-to-day job instruction of the operatives. There are four main points to consider in job instruction.

Preparation

The instructor must prepare both himself and the trainee. He will prepare himself by being sure that all materials, plant and equipment required are available and ready for the instruction and that he is quite clear himself exactly what he is going to teach. Teaching can be ruined if it does not go off smoothly. The trainee should be put at his ease and not made to feel inadequate. The technique of 'reminding' rather than 'teaching' usually has good results with the operative who is new to the company but has had some experience. The instructor should also try to create interest in the job, pointing out whenever possible the place it has in the project as a whole. He should avoid telling a trainee, 'there's really nothing to this job – anybody could do it' as the last thing this will do is create interest.

Presentation

Having carried out the preparation the instructor should then tell, and whenever possible show, the trainee how the job is done, first at the speed it is normally done and then slowly, step by step, explaining the different points. Whilst this is going on it will be most sensible for the trainee to watch from the approximate position of the operator, standing by the side of or looking over the shoulder of the instructor. If he watches facing the instructor he will see the job backwards and not adjust himself to the new situation quickly.

Application

Now the trainee should take over completing the operation slowly and explaining what he is doing. If he falters the instructor should prompt him and if the trainee is still unable to complete the job after prompting then the instructor should 'present' the operation again. When the trainee can complete the job correctly he should be left to proceed.

Testing and follow-up

The final step in the instruction process is for the instructor to check after a short while that the trainee is still performing the operation properly and also to maintain regular follow-up checks to ensure that the performance is maintained and, where possible, improved.

Hazard identification and avoidance

The employer, through the supervisor, has a responsibility to inform employees of any potential hazards, not only in the context of physical work, but also those that could arise from materials that are either being used or could be encountered.

Asbestos immediately comes to mind but there are many other dangerous substances in use on construction sites; highly flammable liquids, fuels, solvents, paints, adhesives, etc; toxic chemicals, acids, degreasants, etc; radiation hazards and explosives. The supervisor should be sure that employees have an appreciation of the possible dangers that can arise. Fuller details are given in particular chapters, as is information on the specific legal duty to train those employees required to mount or change abrasive wheels or discs, but this duty to inform the workforce needs to be emphasized in this chapter dealing with instruction.

Other training

Ideally, safety training should be incorporated in job instruction but there are occasions when short specialized safety courses are useful. For the site operatives the occasional canteen film or slide show and talk given by a visiting expert, safety officer or company plant manager, for example, can be very worthwhile in the business of changing attitudes. The prophet is seldom heeded in his own land and a visitor explaining the need for, say, care with tying ladders often is more effective than the regular cries of the supervisor.

Site supervision, the supervisor himself and his junior supervisors, will really benefit from attending an external course. There are several construction safety organizations offering suitable training of this sort. The Construction Health and Safety Group is probably the leader in the field and caters for anyone within striking distance of London, whilst similar centres operate from Leeds and Liverpool.

All are controlled by practical construction safety professionals and attendance at one of the offered courses would give any supervisor, or prospective supervisor, the right sort of knowledge and enthusiasm that he needs to control work successfully on his site in a safe manner.

Part C

Hazards and safe working methods

5

Demolition and site clearance

Demolition is a skilled and sometimes dangerous operation which requires special procedures if it is to be carried out satisfactorily. Regrettably not every site manager considers the demolition content of a job important and some allow inexperienced operatives to tackle jobs of demolition well beyond their capability. This haphazard approach regularly leads to serious and sometimes fatal accidents and the importance of proper planning for and execution of the work cannot be overemphasized.

Preliminary planning

Site survey

Before any demolition is undertaken it is vital that a detailed survey of the building or structure be carried out. Particular attention should be given to any adjoining property which could be affected by the demolition and the practice of taking photographs of such property is to be recommended. Many a claim for damage has been cut short by the production of photographic evidence of this sort. Work in the vicinity of computer installations or hospitals will need particular care, as falling rubble may set up ground or structural tremors which could affect sensitive equipment.

Walls, balconies, staircases

A careful check should be made of the condition of the walls of the structure to be demolished so as to avoid possibility of premature collapse. Check whether walls are loadbearing and whether cross walls, or party walls, are bonded into the main structure. Particular attention should be given to balconies and staircases of a cantilevered construction and generally these are best removed before the main demolition starts.

Special buildings

Many buildings rely for their stability on the provision of cladding or decking which stiffens the main structural framework. Unplanned demolition of cladding or decking can easily and quickly remove the stability of the whole structure and it is vital that a full investigation be made of this type of operation before work commences.

Details of the construction should be sought from either the original designers, the contractors who erected the structure or the responsible Authority who approved the original plans, and particular care should be taken when the presence of prestressed members is indicated. The hazards of prestressed concrete are dealt with more fully later in this chapter.

Basement areas

Many basements and cellars extend under public footpaths and careful checks and measurements should be made before any demolition is allowed.

Existing services

Consideration must be given to those main supplies and services which may cross or enter the site area and arrangements must be made for appropriate diversions or protections.

Electricity cables and gas supplies obviously present the most serious safety hazards but site managers should not lose sight of the cost element of disrupting other services such as water, telephone cables, etc.

Supervision

Competent supervision is most important in demolition work and the site manager should satisfy himself that the demolition foreman is experienced in the type of work concerned and is able to control the operation in a properly planned sequence of work.

Regulations

This is, in fact, a requirement of Part X of the Construction (General Provisions) Regulations 1961 which requires competent and experienced supervision and additionally specifies certain demolition operations that must *only* be undertaken under the immediate supervision of a foreman with adequate experience of that kind of work, or by experienced demolition operatives who have been instructed in the work method by the foreman.

These operations are:

1. The demolition of the whole or part of a structure, except where there is no risk of collapse.
2. The demolition of a structure where there is a special risk of collapse.
3. The cutting of reinforced concrete, steelwork or ironwork forming part of the structure to be demolished.

Study of the above will indicate that very few demolition operations fall outside these definitions – a point that should always be in the site manager's mind.

Safety of operations

Regulations

Part X of the Construction (General Provisions) Regulations 1961 has already been referred to and there are, in addition, many other Acts and Regulations that may apply, depending on the particular demolition operation being tackled. A list of these is given in Table 5.1.

Protective equipment

Very few demolition operations exist where the wearing of helmets, safety footwear, gloves and goggles is not necessary and every effort should be made to encourage operatives to wear these items of protective equipment. Special operations will require additional precautions to be taken and the wearing of suitable respirators and special clothing must always be considered in those demolition areas where toxic chemicals, fumes or asbestos are to be found.

Plant safety

It is unfortunate but often true that items of plant provided by the demolition contractor for use in his work are not of the highest standard. It is, of course, a rough, heavy operation and one in which plant may easily be damaged. There is, however, no excuse or reason for the site manager to allow the use of unsafe plant on the site and all the usual inspections and certifications apply and should be rigidly enforced.

Table 5.1 List of relevant Acts and Regulations

Clean Air Act 1969
Control of Pollution Act 1974
Explosives Act 1975 and 1923
Factories Act 1961
  The Asbestos Regulations 1969
  The Construction (General Provisions) Regulations 1961
  The Construction (Lifting Operations) Regulations 1961
  The Construction (Health & Welfare) Regulations 1966
  The Construction (Working Places) Regulations 1966
  The Electricity Regulations 1908 and 1944
  The Health and Safety (First Aid) Regulations, 1981
  The Highly Flammable Liquids and Liquefied Petroleum Gases Regulations 1972
  The Lead Paint Regulations 1927
  The Protection of Eyes Regulations 1974
Greater London Council (General Powers) Act 1966
Health and Safety at Work etc. Act 1974
Highways Act 1959
London Building Acts (amendment) Act 1939
Public Health (London) Act 1936
Public Utilities Street Works Act 1950
Town and Country Planning Act 1974
Water Act 1945

Applicable to Scotland only

Building (Scotland) Act 1959
The Building Operations (Scotland) Regulations 1963
The Building (Scotland) Act 1959, (Procedure) Regulations 1964
The Building (Scotland) (Forms) Regulations 1964

Note: There may also be local authority by-laws to consider and further enquiries are advised in this respect.

Safety of the public and protection of adjacent property

Consideration must always be given to preventing injury to any third parties, such as passers-by or residents or workers in adjacent property, and also to providing protection for adjacent property.

Adequate fencing or hoarding must be provided to ensure that the site is closed against entry both while work is proceeding and, just as important, when work has finished for the day. Care should also be taken to ensure that the structure is left in a safe condition at the end of the working day, as no fence has yet been invented that will keep out determined schoolchildren. It must also be appreciated that demolition methods using mechanical means, and in particular the demolition ball, can cause debris to fly a considerable distance and this method should not be permitted close to the site boundary. Dust screens should be provided when working in built-up areas and regular hosing of the work area is necessary in dry weather.

Bonfires and smoke

Many local authorities will prohibit the lighting of bonfires on demolition sites but if they are permitted, then they must be carefully controlled. Always ensure that any fires are properly extinguished well before the operatives leave the site and that materials that are burned are not likely to give off toxic or offensive smoke. Remember that any smoke in sufficient quantity can present a visibility difficulty and consequent hazard to passing vehicles.

Pedestrian and vehicular access

Whilst it is essential that third parties are excluded from the work area it may also be necessary, in some cases, to provide access through or near to the site for residents or adjoining property owners. Proper regard must be paid to any requirements of this sort and, if necessary, protective covering or fans should be erected over the access ways. These coverings must be constructed of suitable materials and of sufficient strength for the purpose. Consideration must also be given to the need for providing temporary lighting for any access ways provided.

Demolition methods

Hand demolition

Wherever possible demolition contractors will, understandably, use mechanical methods in their demolition operations but there are many situations where hand demolition is the only acceptable method. This method, where the operatives demolish the structure using hand-held tools can, unless carefully planned, be more hazardous to the operatives than any other method and there are several important rules that must be applied.

1. Demolition should be progressive and carried out in the reverse order to that of construction.
2. If the work cannot be safely done from a part of the structure then a suitable scaffold platform should be provided. Scaffolds for demolition work are somewhat special as they need to be so constructed that they can be progress-

ively dismantled as the demolition work proceeds. The normal good scaffolding practice of staggering the position of joints in the standards is best not followed when erecting a demolition scaffold as this can present difficulties when a progressive dismantling of the scaffold is required.

3. Where debris is to be allowed to fall internally, i.e. within the walls of the building or structure, care must be taken to ensure that there are sufficient openings that are clear of joists etc. so that the debris may fall safely without being deflected or building up on intermediate floors. Attention must also be given to the possibility of lateral pressure building up against the walls of lower storeys and regular removal of debris is necessary. Precautions must also be taken against flying debris by sealing off wall openings in the immediate area of the debris fall.
4. If the debris must be dropped externally, hazards of containment are greater. This method should not in any case be employed if the public highway or nearby property is within 6 m from the area of fall and, with buildings or structures more than 12 m tall, then no public highway or property should be closer than half the height from which the debris will fall. The alternative to free fall of debris is to use chutes or skips.
5. The use of an experienced reliable look-out man is to be recommended whenever debris is being dropped in this way.

Mechanical demolition

Demolition ball

Probably the most common method of mechanical demolition is the demolition ball and there are three techniques that can be used:

1. Vertical dropping of the ball from a fixed jib position.
2. Swinging the ball in line with the jib.
3. Swinging the ball by slewing the jib.

There are several important points of which the site manager should be aware:

1. When the demolition ball is attached to a normal duty mobile crane then ONLY the vertical drop technique must

be used. Heavy duty machines are necessary for the other techniques.

2. The hoist rope of the crane, the anti-spin device, which should always be fitted in conjunction with the demolition ball and the attachment of the ball to the hoist rope should be inspected at least twice daily.
3. Operators must be experienced and skilled not only in normal crane operations but in the techniques of demolition.

The pusher arm

This is a method which involves a machine fitted with a pusher arm which exerts a thrust in the horizontal plane. This method has the limitation that it can only be used on walls or structures that are of a height to suit the machine, as the point where the pushing arm is applied should not be more than 600 mm below the top of the wall.

Deliberate collapse

This is a method whereby key structural members are removed causing the whole building or structure to collapse. The method adopted should ensure that the demolition is completed in one operation and that no parts of the structure are left in an unstable condition, thus causing a possible hazard to operatives required to finish off the job.

If the planned method involves the use of a wire rope for pulling out weakened structural members, several important points should be considered:

1. The rope should be a steel wire rope of a size adequate for the purpose and in no case smaller than 38 mm in circumference.
2. When pulling begins no persons should be forward of the pulling machine or on either side of the rope within a distance of three-quarters of the distance between the pulling machine and the object to be pulled.

Use of explosives

Certainly the use of explosives results in the most spectacular demolition, but this is even more of a specialist operation than the other methods described and the services of an experienced consultant should be sought before any decisions are made on the use of this method.

Table 5.2 Guide to recommended demolition methods

A denotes hand demolition
B denotes use of pusher arm
C denotes deliberate collapse method
D denotes use of demolition ball
E denotes demolition by explosives

Type of structure	Type of construction	Location of site			
		Detached building isolated site	Detached building confined site	Attached building isolated site	Attached building confined site
Small and medium two-storey buildings	Load-bearing walls	ABCD	ABD	ABD	AD
Large buildings, three storeys and over	Load-bearing walls	ABDE	ABDE	ABDE	AD
	Load-bearing walls with W.I and C.I members	ABDE	AE	AE	AE
Framed structures	Structural steel	ACE	AE	AE	A
	Reinforced concrete	ADE	ADE	ADE	AE
	Pre-stressed concrete see special section in text				
	Composite structures, steel and RC	ADE	ADE	ADE	A
	Timber	ABCDE	ABD	ABDE	ABD
Cantilevers (balconies and stair-cases etc.)	Independent of a framed structure	ADE	ADE	ADE	ADE
Bridges		ABCDE	ABCDE	AE	AE
Masonry arches		ACDE	ACDE	ACDE	ACDE
Chimneys	Brick or masonry	ACDE	A	ACDE	A
	Steel	ACE	A	AE	A
	Reinforced concrete	ADE	A	ADE	A
	Plastics reinforced	A	A	A	A
Spires		ACDE	A	A	A
Pylons and masts		ACE	A	A	A

(Adapted from BSI Code of Practice)

Choosing the method The demolition method chosen will depend on three main considerations: the type of structure, the type of construction and the location of the site. A guide to the type of demolition recommended for various buildings and structures is given in Table 5.2.

Special problems when demolishing prestressed concrete

The demolition of a building or structure containing prestressed members can create particular hazards and a basic understanding of the problems is necessary. There are two basic methods of prestressing:

1. Pre-tensioning, when steel wires or tendons are tensioned before the concrete is cast around them and the concrete bonds with the steel.
2. Post-tensioning, when the wires or tendons run through voids pre-formed in the concrete and are tensioned after the concrete has reached its design strength. Anchorages are, of course, provided to sustain the stress in the tendons and in most cases the tendons are subsequently grouted in.

Pre-tensioned units are generally produced in factory conditions whereas post-tensioning techniques are usually applied to units which are cast *in situ*.

It must be appreciated that a rapid release of the tension in any prestressed member by removing the concrete surrounding the tendons, by cutting through the steel or by removing anchorages can lead to a sudden collapse or failure of the member. There is also a distinct possibility that, under such circumstances, the released tendons or anchorages can become missiles.

It becomes vital therefore that the pre-demolition survey picks up any likelihood of the existence of any form of prestressing and that any such members are demolished in accordance with the advice given by a Chartered Engineer experienced in this type of construction.

6

Excavations and earthworks

Many safety specialists take the view that a trench excavation in 'running sand' causes them, the safety specialists, least bother – explaining that this type of soil makes first-class support work necessary to enable *any* work to proceed and that accordingly safety looks after itself. There is undoubtedly more than a grain of truth in this opinion as it is generally accepted that most collapses in excavations occur because of a lack of suitable support materials in so-called 'good' ground.

Accidents following the collapse of excavations usually involve serious injuries and regularly result in fatalities. One has only to be reminded that in round figures a cubic yard of earth weighs at least a ton to appreciate the effects of even a small fall of earth.

It becomes obvious then that excavation work must be treated very seriously and it would be sensible in the first instance to look at the legal requirements of the regulations.

Requirements of Regulations

Apart from the general duty to provide a safe place of work imposed by the Health and Safety at Work etc. Act, excavations are covered in more detail by Regulations 8–14 of the Construction (General Provisions) Regulations 1961. The regulations should be studied in their entirety but a simplified interpretation is given below.

Regulation 8 – Supply and use of timber

In excavations more than 4 ft (1.2 m) deep, where there is a risk of collapse, proper support work must be provided and it should be noted that protection must also be given to those operatives who are positioning and fixing the support work. The regulation does not apply if the sides of the excavation are sloped back to a safe angle so that there is no risk of a collapse or of men being struck by material falling from a height of more than 4 ft (1.2 m).

Regulation 9 – Inspections and examinations

Inspections must be made every day that men are working in an excavation more than 4 ft (1.2 m) deep, and, at the beginning of every shift, of the working ends of trenches more than 6 ft 6 in. (2 m) deep and of the base of any shaft.

Whilst no record need be kept of *these* inspections, there is also a requirement for a more thorough examination of excavations to be carried out:

(a) after the firing of any explosive charge used in the excavation process;
(b) after any damage to support materials or any fall of earth or material;
(c) *in any case, every seven days* and a record of *these* examinations must be kept in the official register for the purpose, Form 91, Part 1, Section B.

The examinations must be made by a person suitably experienced in excavations, quite often the general foreman, and the register entries must be signed by him.

Regulation 10 – Supervision of work

Materials must be inspected before use and any defective material must be rejected.

Wherever practicable only experienced operatives should be engaged in erecting, altering or dismantling support work and they must be supervised by a competent person. The work must be properly constructed, with particular care taken to ensure that struts and braces cannot be accidentally displaced and, once constructed, it must be maintained in good order.

Regulation 11 – Means of egress in case of flooding

This regulation makes special provision for excavations where there is a risk of flooding and requires suitable escape ladders to be provided. The provision of ladders is, of course, essential in excavations irrespective of any possible flooding risk.

Regulation 12 – Excavations likely to affect adjacent structure

Consideration must be given to the possibility of any adjacent structures being affected by an excavation and provision made for shoring or other methods of support.

Regulation 13 – Fencing of excavations

Excavations of any sort, more than 6 ft 6 in. (2 m) deep, near which employees work or pass must be protected by guardrails or barriers or must be securely covered. The barriers may be removed temporarily for access of men, plant or materials but must be replaced as quickly as possible. Plastic bunting is not acceptable as a physical barrier but may be used in certain circumstances providing it is set back some 1 m from the risk.

Regulation 14 – Safeguarding edges of excavations

Materials and plant must be kept well away from the edges of all excavations to avoid creating an overburden and a possible collapse of the sides and also to remove the risk of men and machines falling in (see Fig 6.1).

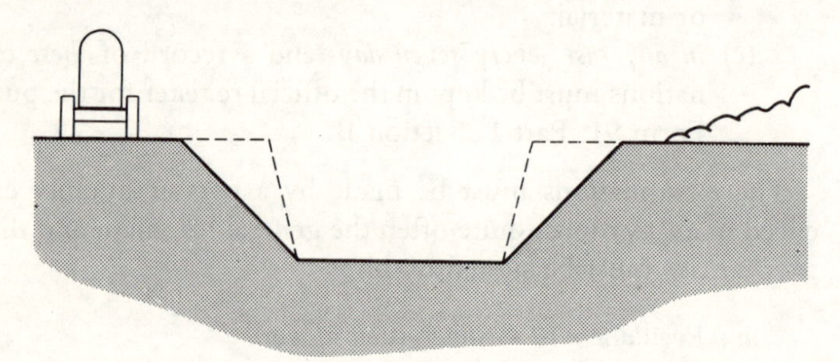

Fig. 6.1 Preventing overburden. Sides of excavations cut well back. Spoil heaps and plant kept clear of sides of excavations.

These few regulations have all been formulated as a result of statistical evidence showing that failure to meet these standards has led to regularly repeated accidents. The regulations do not deal with support methods and the reader is recommended to refer to the British Standard Code of Practice CP 6031: *Earthworks* for detailed advice, although some of the more common types will be described later in this chapter.

Soil characteristics and necessary support

There would be a great improvement in the standard of excavation safety if it was always accepted that every excavation of more than 4 ft (1.2 m) depth needed support unless the sides were sloped back to a safe angle of repose.

If this was accepted, then the type of support that was suitable and adequate would depend on the ground to be excavated, and there are basically three major types of soil to be considered: rocks, cohesive soils and non-cohesive soils.

Rocks

There is a wide variety of types in this category ranging from hard sound rocks, through the limestones and sandstones to the soft rocks and chalk. It is dangerous to assume that excavations in rock are necessarily safe, as all rock formations are naturally split into blocks of varying size running in layers and separated in many cases by thin layers of water or clay. If these layers run at an angle to the horizontal then the blocks can slide into an excavation or trench that is opened up, assisted by the lubricating role of the water/clay layers. Figure 6.2 illustrates this.

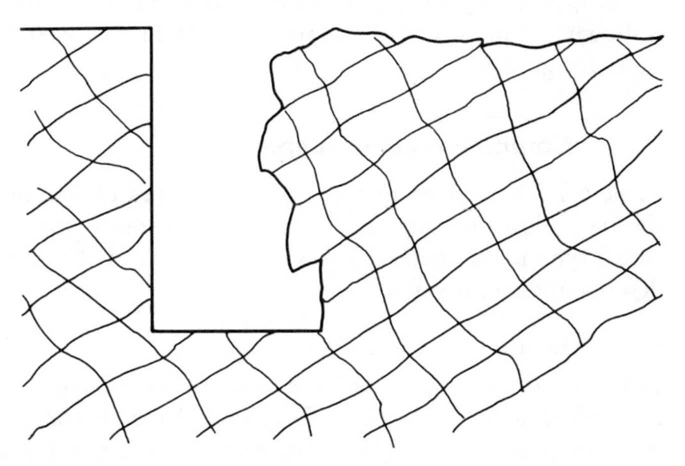

Fig. 6.2 Excavation in rock. Trench side collapses in excavation in steeply dipping rock. Blocks of rock slide along bedding planes.

The very action of the excavation process, use of rock breakers and drills, explosives, etc., can obviously contribute to loosening the blocks of rock near the excavation. All excavations in rock need to be carefully executed, with particular attention being paid to inspections of the exposed surfaces and the possibility of weather conditions loosening face materials.

Cohesive soils

This collective term includes the clay type of soil, from the very stiff and hard clay to the soft clay and is the most treacherous with which to deal. This particularly applies to excavations in the firmer clays, and the apparently stable appearance of the dig has many times wrongly encouraged workers to engage in risk-taking, sometimes with fatal results. The effects of time and weather conditions on newly exposed faces of excavations can be quite startling, with rapid deterioration of the stability of the face, and even in shallow trenches some minimal timbering will be necessary.

Non-cohesive soils

Excavations in gravels and sands will inevitably need support or sloping back of the sides, and, while sloping of the sides is a very safe method of working, it can require, with non-cohesive soils, a great deal of working space and is often economically unacceptable.

The type of support required for excavations with vertical sides is shown in Table 6.1 which is reproduced from BS CP 6031: *Earthworks* for guidance.

Common types of support

Table 6.1 has referred to three common types of support, open sheeting, close sheeting and sheet piling. A fourth common and useful type is the method known as 'pinchers' and I will briefly describe these four types.

Pinchers

In harder ground, where the need for temporary support is less, it is still sensible to provide a form of skeleton sheeting. Pairs of supporting sheets set at about 1 m apart, or as required, and strutted by one or two struts can be used sensibly in shallow trenches. The sheets should be toed-in to the trench bottom to give additional stability. This method does not make use of walings, which are the horizontal timbers used in other methods to support a row of vertical sheets, but struts are butted directly against opposite sheets (see Fig. 6.3).

Open sheeting

For slightly less firm grounds a system of open sheeting can be employed. This method, as illustrated, does make use of walings but the vertical sheets are not kept closely together. This results in a saving in the number of sheets required but herein lies a danger

Table 6.1 Support required for excavations with vertical sides in uniform ground
A Indicates that no support is required
B Indicates that open sheeting should be employed
C Indicates that close sheeting or sheet piling should be employed.

Type of soil	Depth of excavation		
	Up to 1.5 m (shallow)	1.5 m to 4 m (medium)	Over 4 m (deep)
Soft peat	C	C	C
Firm peat	A	C	C
Soft clay	C	C	C
Firm and stiff clays	A*	A*	C
Loose gravels and sands	C	C	C
Slightly cemented gravels and sands	A	B	C
Compact gravels and sands with or without clay binder	A	B	C
All gravels and sands below water table	C	C	C
Fissured or heavily jointed rocks	A*	A*	B
Sound rock	A	A	A

* Open or close sheeting or sheet pilling may be required if site conditions are unfavourable.

and the supervisor should check any tendency on the part of his ground workers to spread the sheets too far apart (see Fig. 6.4).

Close sheeting

As its name suggests, the close sheeting method provides a complete wall of sheets supported by walings which are are in turn cross strutted, and is used where ground conditions require this treatment (see Fig. 6.5).

Sheet piling

The end result of a sheet piled trench or excavation is similar to that of the close sheeted system except that the sheet piles are usu-

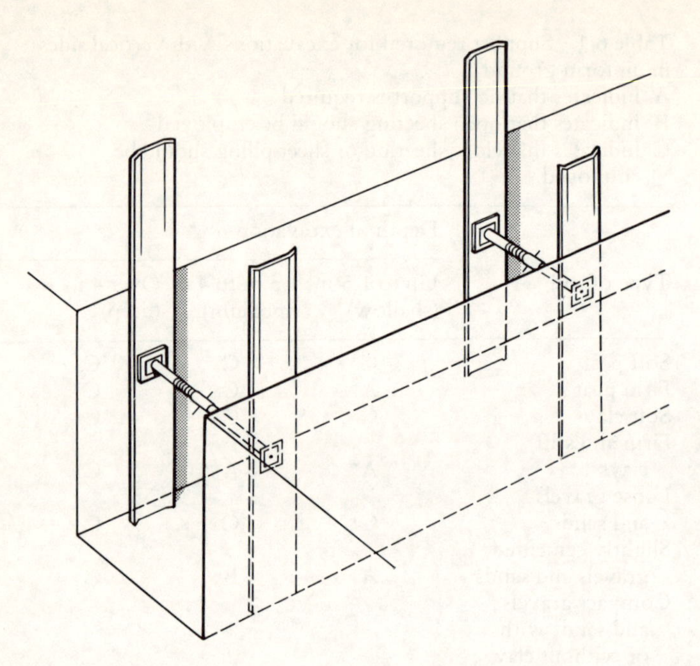

Fig. 6.3 Pinchers. *Note*: 'toe-in' of steel sheets and timber packing between sheets and props.

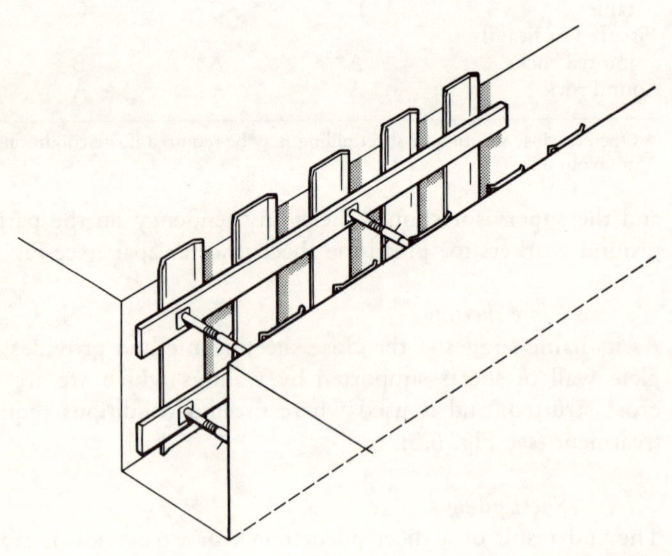

Fig. 6.4 Open sheeting. *Note*: use of walings to support run of sheets using adjustable struts.

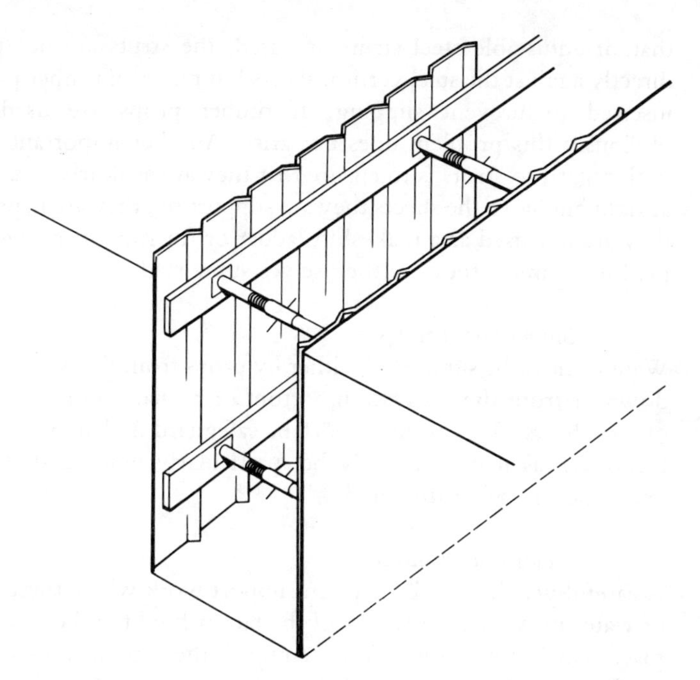

Fig. 6.5 Close sheeting.

ally of heavier steel and the finished job much stronger in consequence. The main difference is that sheet piling is normally driven by an air operated hammer, or drop hammer *before* any excavation takes place. Excavation begins when the piles are to a depth below the depth of the dig and provided with walings and struts as the excavation progresses.

Safety features of common support methods.

These descriptions of the common methods of support are necessarily brief and readers needing more information on the actual design and methods of support work are again referred to BS CP 6031. The important points on which the supervisor should concentrate are those safety features of support work which are sometimes disregarded by operatives or not appreciated by them as being possibly dangerous.

Strutting for pinchers

As no walings are used in this method care should be taken to see that, if adjustable steel struts are used, the struts are not placed directly against the steel vertical sheets but pieces of timber packing inserted to prevent slipping. If timber props are used then obviously this problem does not arise. Another important aspect with regard to struts is to ensure that they are as nearly as possible at right angles to the sheets they are supporting; they are kept tight; they are not used as a makeshift ladder or for supporting working platforms unless they are themselves supported.

Supporting walings

Walings must be supported, either by props from the waling lower down or from the excavation bottom or by hangers from the top of the sheets. This is an important safeguard and must never be forgotten, as it can so easily be, particularly when a sheet piled excavation is being timbered.

Voids behind sheets

A careful watch must be kept on support work where there is a lot of water movement. Erosion of the face behind the sheets can take place which can result in a loosening of the supporting struts and a subsequent failure of the sheeting.

Obstructions

Every care should be taken to locate and remove any obstructions before commencing to drive sheet piles. The use of an electronic pipe and cable locator is strongly recommended. You may have an assurance from the statutory authorities that no services are present but public records are not always absolutely accurate and a double check is worthwhile.

Access, egress and protection of excavations

Trench ladders of adequate length and strength must be provided to all excavations. There must be at least 1 m hand hold above the landing point of any ladder and some means of securing the ladder, to prevent it slipping sideways, must be provided. In large excavations and long trench runs a sufficient number of ladders is required and a rule of thumb for this is one ladder every 15 m.

The provision of fencing has been covered previously in this chapter but equally important is the need for barriers and stop blocks positioned to prevent plant and transport driving or reversing into excavations. Be sure that the barriers you provide are large enough to do the job. Large wheeled dump trucks can easily overrun an inadequate block. Items of plant must be carefully positioned not to put an overburden on the side of the excavation and similar care must be taken with the placing of spoil heaps. A useful guide is to keep spoil back from the excavation edge at a distance equal to the depth of the excavation.

Lighting of and in excavations

The need for warning lights around excavations which are near or on the public highway must be emphasized. Reflective signs are not in themselves sufficient; they must be supplemented by amber lights during the hours of darkness.

Lighting in excavations may be required in deeper excavations, especially in the winter months. A reduced voltage of 110 V should be used for the supply and extreme care taken to position the cables and lamps in safe positions.

Ventilation

It must be remembered that many flammable and toxic gases are heavier than air and that they find a natural receptacle in excavations. Great care must be taken when positioning plant to see that exhaust gases are not fed into excavations. Liquefied petroleum gases are also notorious for this danger and a careful watch must be kept on gas cylinders and possible leaks of gas.

Additionally the presence of natural gases may be encountered, gases such as methane and sulphur dioxide, and the supervisor must regularly impress on his ground workers the need for a constant look-out for a foulness of the air in any excavation situation.

The remedy for many of the examples is to prevent the gases and fumes from entering the excavation but in some cases it will be necessary to provide forced ventilation by introducing clean air in sufficient quantities to dissipate the fumes and gases. In such circumstances you must establish what the problem gas is and regularly test the atmosphere in the excavation to ensure that the ventilation method is proving to be effective.

Proprietary methods for supporting trenches

In very bad ground conditions, where the exposed faces of the trench excavation will not stand long enough to allow support work to be positioned in the usual way and where for some reason sheet piling is not desirable, it may be sensible and necessary to consider one of the proprietary methods of trench support. These vary from fairly lightweight pairs of support sheets, hydraulically strutted (which can be very quickly dropped into a trench and pumped into position either to provide a part of the support or to act as a temporary haven for the operative placing more traditional support materials) to fairly large trench size shields which are pulled along behind the digging excavator, providing a very secure place of work for operatives.

All the methods tend to appear expensive but can prove very effective in time saving, undoubtedly improve the safety aspect and should always be considered when ground conditions are so bad that sheet piling is the only other alternative.

7

Scaffolding and means of access

The provision of adequate scaffolding, which is properly maintained, is vital on construction sites and the erection of scaffolding by inexperienced workmen, however small the firm or site may be, must be rejected. Past experience has shown that ill-planned, badly erected scaffolds can lead to serious accidents and this is no less true of the many types of frame scaffolding now in use. This type of scaffolding may have advantages in that it is generally easier to erect, having several tubular members and couplers combined in one frame, but it is this very 'easiness' which can lead to the inexperienced worker 'trying it out' and sometimes discovering to his cost that it is not as easy as it looks. More of this later.

Regulations

The regulations which govern this aspect of safety are contained in the Construction (Working Places) Regulations 1966. This is a fairly long code, running into thirty-nine reasonably detailed regulations, and is best studied as a separate document. What I plan to do is to make reference to the regulations when describing erection principles and safety measures required, but it is worth while quoting one of the early regulations in the code.

Regulation 7, simply interpreted, says that: 'Scaffolds must be provided for all work which cannot safely be done from the ground or from part of the permanent structure. Ladders may be used but only for light work which can be done with one hand.'

This really places the obligation on the employer, i.e. the supervisor, to make sure that no workers are asked to work above ground level unless on a safe and secure footing.

For the supervisor requiring a more detailed understanding of scaffolding a study of the British Standard Code of Practice CP 97:

Metal Scaffolding is recommended in addition to the statutory regulations.

Scaffolders record scheme

In an attempt to improve the standard of expertise of scaffolders the employers federations and the unions concerned have drawn up a scheme which requires scaffolders to be trained to an agreed standard (trainee, basic and advanced) provides for a system of training record cards for each individual 'scaffolder' and requires employers, who are party to the working rules of the NJCBI and CECCB, to employ only trained operatives who hold appropriate record cards for the work in question. The working rule in fact provides that an operative who has not attained prescribed levels of training and experience in scaffolding of a given kind must not be employed on them unless:

1. Under adequate supervision.
2. Working with an operative who has the required training and experience.
3. Erecting, altering or dismantling simple access scaffolding with a working platform no higher than 5 m.

So it will be seen that employers who are party to the working rule should ensure that all scaffolds over 5 m high are erected only by trained operatives.

It will be appreciated that many small sites, housing and low-rise buildings fall outside the scope of the scheme and, whilst it is obviously desirable even on basic scaffolding that only trained scaffolders are used, this may not be the case in practice. It should also be appreciated that there will always be loop-holes in any record scheme, that there will still be the odd 'cowboy', that in any case the supervisor's responsibility still remains and accordingly he should acquire a basic knowledge of the scaffolding skills.

Scaffolding terms

It is important that we all speak the same language and listed below are some of the more common scaffolding terms.

Tubular members

Brace A tube incorporated diagonally across two or more members in a scaffold and fixed to them to afford stability.

Guard rail or handrail A member incorporated in the structure to prevent the fall of an operative.

Ledger A tube spanning horizontally and tying a scaffold longitudinally which may act as a support for putlogs or transoms.

Putlog A tube with a flattened end or other member spanning from a horizontal member to a bearing on the wall of a building and which may be used to support a working platform.

Raker An inclined loadbearing tube.

Reveal tube A tube wedged between two opposing surfaces, e.g. window reveals to form a friction anchor to which the scaffold may be tied.

Standard A vertical or near vertical supporting member.

Tie A member used to tie the scaffold to a structure.

Transom A tube spanning across ledgers to tie a scaffold transversely which may also support a working platform.

Scaffold fittings

Baseplate A plate for distributing the load from a standard or raker.

Coupler, putlog A coupler used for fixing a putlog or transom to a ledger.

Coupler, right angle A coupler other than a putlog coupler, used for connecting two tubes at right angles.

Coupler, swivel A coupler for connecting two tubes at any angle where a right angle coupler cannot be used.

Coupler, sleeve An external fitting for joining two tubes end to end.

Joint pin (or expanding spigot) An internal fitting for joining two tubes end to end.

Reveal pin The fitting used for tightening a tube between two opposing surfaces.

Toeboard clip A clip used for attaching toeboards to scaffolding members.

General

Bay The space between the centre lines of two adjacent standards along the face of the scaffold.

Lift The height from the ground or floor to the lowest ledger, or the vertical distance between adjacent ledgers.

Sole plate A timber or other member of adequate size and suitable quality used to distribute the load from the baseplate to the ground.

Toeboard A board set on edge used to prevent tools, materials or feet from slipping off the platform.

Construction common to all scaffolds

Whilst separate sections will deal with the construction of particular types of scaffolding, there are a number of points that apply generally, and these are given here.

First the scaffolding materials to be used. It is necessary to satisfy yourself that the parts are suitable before using them.

In the case of tube, it should be straight and the ends should be sawn square to the axis of the tube. Severe corrosion can usually be seen by a thinning of the tube wall at the ends which seriously reduces the load bearing capacity. Do not use such tubes and request that they be properly checked by a person technically qualified.

Fittings should be clean and threads lightly oiled. Do not use any coupler that appears to be misshapen.

Scaffold boards should be clean, free of nails, not split or badly warped and the steel banding should not be torn or jagged.

Ladders may be varnished or treated with preservative, but must not be painted. Check that all rungs are sound and properly wedged. See that stiles are not warped, cracked or splintered.

A scaffold must not be erected upon an unprepared foundation. If soil is the base, it must be well rammed and timber soleplates at least 9 in. (230 mm) wide and (1½ in.) (38 mm) thick laid upon it so that there is no air space between timber and ground.

Standards should be pitched on baseplates and any joints in the standard should occur just above the ledger. These joints should be staggered in adjacent standards so that they do not occur in the same lift.

Ledgers should be horizontal, placed inside the standards and clamped to them with right-angle couplers. Joints should be staggered so that in adjacent ledgers they do not occur in the same bay. It is strongly recommended that only *sleeve* couplers are used at ledger joints.

Decking will generally be completed with 9 in. (230 mm) wide × 1½ in. (38 mm) thick boards of varying lengths and each board should have at least three supports. Boards are normally butted, but may be lapped if bevel pieces are fitted or other measures taken to prevent tripping. A 9 in. (230 mm) board should extend between 2 in. (51 mm) and 6 in. (152 mm) beyond its end support.

Guardrails must be fitted at all working platforms between 3 ft (0.914 m) and 3 ft 9 in. (1.143 m) above the platform level. Toeboards and barriers must also be fitted at least 6 in. (152 mm) high, but in any case high enough to prevent material falling from the platform. The space between toeboards and guardrails must not exceed 30 in. (0.762 m). Ladders must stand on a firm and even base and must be secured so that they cannot move from top or bottom. Ladders must extend at least 3 ft 6 in. (1.066 m) beyond the landing place.

Finally, unless properly designed to stand on their own, all scaffolds must be sufficiently and effectively anchored to the building by ties. These ties are essential to ensure stability of the scaffold, and this point cannot be over-emphasized.

Independent tied scaffolds

Do not be misled by the word 'independent'. These scaffolds consist of a double row of standards connected together longitudinally with ledgers and with transoms at right angles to the ledgers.

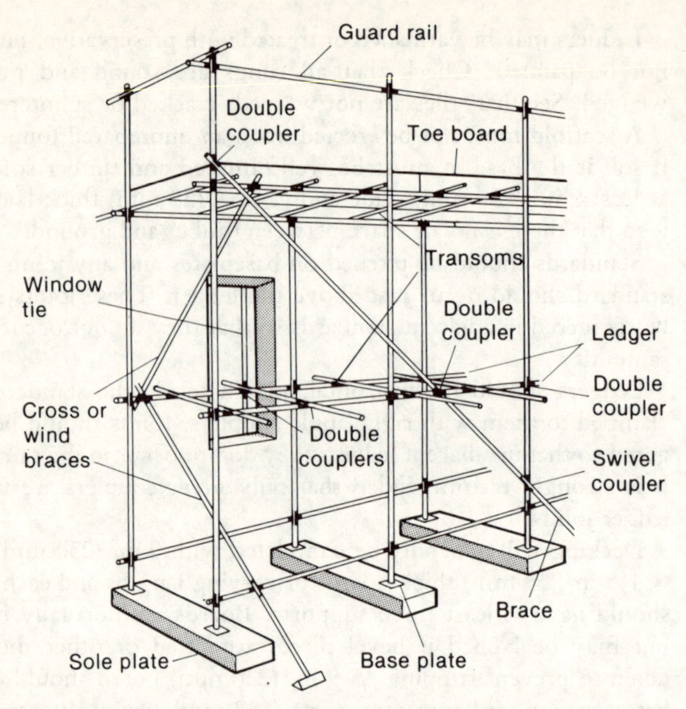

Fig. 7.1 Independent tied scaffold. *NB*. This sketch of an independent tied scaffold is given only as a guide to the position of the various members. It is not to scale and no attempt has been made to reproduce the shape or dimensions of the scaffold fittings.

Bracing and tying are essential for stability (see Fig. 7.1).

The inner row of standards should be placed as close as possible to the building face. To avoid projections such as cornices, these standards may be placed about 13 in. (330 mm) from the wall provided a single board is placed between the wall and the inner standard on transoms that extend to within 2 in. (51 mm) of the wall.

The outer row of standards will be approximately 3 ft 4 in. (1.016 m) from the inner row to allow for four boards. For general purpose scaffolds, if no board is placed at the building face, the outer standards may be placed 4 ft 1 in. (1.244 m) from the inner row, to allow for five boards.

Generally ledgers will be vertically spaced at 2 m centres, to give headroom at all platforms. The first pair of ledgers, however, may be up to a maximum height of 8 ft 6 in. (2.59 m).

Transoms should be placed on the ledgers at not more than 4 ft

(1.219 m) centres. In each bay, one transom should be fixed within 12 in. (305 mm) of a standard with right angle couplers to the standard or ledger and must remain in position. Intermediate transoms may be secured to the ledgers with putlog or right angle couplers and may be removed if not required to support scaffold boards.

Diagonal bracing at right angles to the building at alternate pairs of standards is necessary for the full height of the scaffold. These braces should be fixed to the ledgers with right angle couplers as close to the standards as possible. Where right-angle couplers cannot be used, a swivel coupler may be employed to fix the brace to the standard.

Longitudinal or facade bracing to the full height of the scaffold is necessary spaced not more than 100 ft (30.48 m) apart along the scaffold length. Braces should be fixed at between 30° and 45° to the horizontal, with right-angle couplers to those transoms placed adjacent to the standards. Where right angle couplers cannot be introduced, connection to the standards may be made with swivel couplers.

Longitudinal bracing may alternatively be provided of a zigzag pattern to the full height of the scaffold in the end bays and not more than 100 ft (30.48 m) apart along the scaffold length. The number of braces required may be more accurately determined by fixing one brace for every ten standards.

Joints in braces should be made with sleeve or parallel couplers because of the possibility of tension stresses.

The permitted loading for a general purpose independent tied scaffold, i.e. one with standards set at 7 ft (2.133 m), is 37 lb/ft^2 (180 kg/m^2) on up to four working lifts.

Importance of ties

It is essential that all independent tied scaffolds are securely tied to the building throughout their length and height to prevent movement of the scaffold either towards or away from the building. This should be done by connecting a tie assembly. Where this is not possible, tubes may be securely wedged in openings by reveal pins and clamped to the tie tubes.

To ensure security of reveal ties it is necessary to check frequently for tightness.

If reveal ties are used, they shall not exceed 50 per cent of the total number of ties and the through ties shall be evenly distributed over the scaffold area.

Only right-angle couplers should be used for securing ties. Ties should occur at every other lift or not more than 13 ft (3.96 m) apart vertically, and at not more than 20 ft (6.096 m) intervals along the scaffold.

All decking will generally be 9 in. (230 mm) × 1½ in. (38 mm); scaffold boards and platforms will be four or five boards wide.

Guardrails and toeboards or barriers are required at all working levels.

Independent tied scaffolds erected in accordance with the foregoing directions may be used up to 150 ft (45.72 m) in height. For variations in loading and standard spacing, refer to the Code of Practice.

Temporary rakers are usually employed to brace the scaffold against the ground when setting out. These rakers are replaced by permanent braces when scaffold has been plumbed and tied.

Mobile towers

A mobile tower is formed either with proprietary frames or with tube and fittings and mounted on wheels. It has a single platform, designed to support a distributed load of 145 Kgf/m^2 (30 lb/ft^2), and the standard rules for board supports, guard rails and toe boards all apply.

Bracing of the tower is important and should be provided by fixing diagonal bracing to all four elevations and on plan. Care should be taken to ensure that the castors used with the tower are fitted at the extreme corners and that they cannot fall out when the tower is moved. They must also be provided with an effective wheel brake to be locked when the tower is in the required position. The actual moving of the towers requires great care and positive supervision to ensure that all persons, equipment and material are removed from the platform before moving starts. Under no circumstances must mobile towers be moved by persons on the platform pulling the tower along.

Height limitations

A very simple formula should be used for establishing a safe height for a mobile tower.

When being used externally then the height of the platform should not exceed three times the shorter base dimension. If the

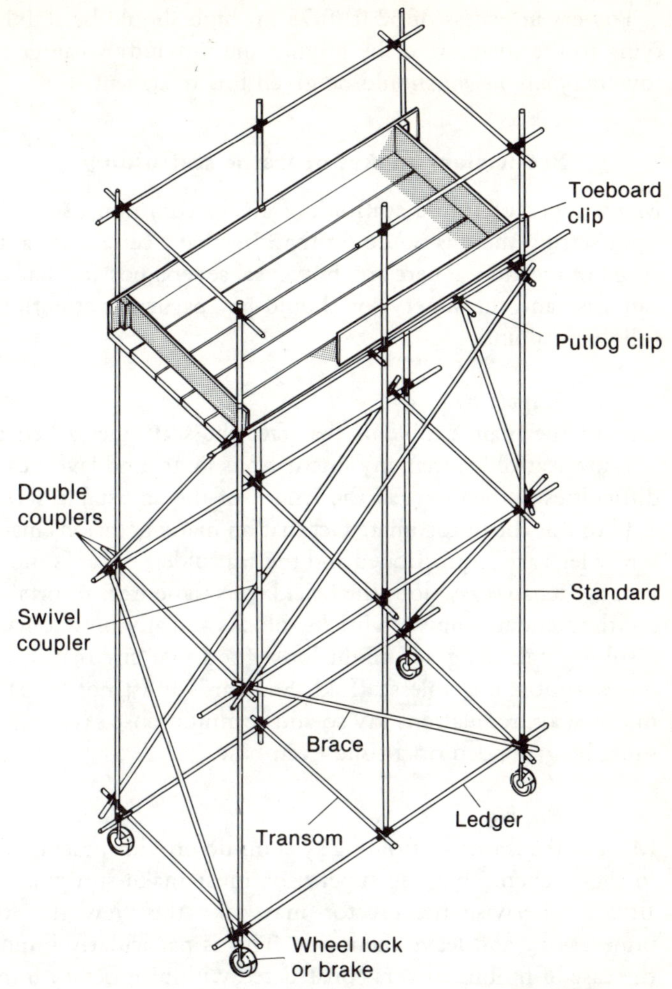

Fig. 7.2 Mobile tower. *NB*. This sketch is not to scale. All couplers should be taken to be double couplers except where indicated.

tower is to be used internally then the platform height can be increased to three-and-a-half times the shorter base dimension. Additionally it is recommended that no tower should have a base dimension shorter than 4 ft (1.219 m).

Additional dimensions at the base can be obtained by fitting outriggers, so increasing the permitted platform height without increasing the overall size of the tower, but again the need for supervision is paramount.

Towers in excess of 32 ft (9.75 m) high should be stabilized by tying to the structure or by guying and in windy weather external towers *of any height* should be given this treatment.

Proprietary makes of frame scaffolding

Many types of frame scaffolding are in common use and in the right setting there is no doubt that they have certain advantages in speed of erection. There are, however, several possible hazards that can arise and the supervisor should pay particular attention to the following points:

Scaffold base

One of the main problems that face the scaffolder asked to erect a frame scaffold is that any fluctuations in ground levels can create difficulties if they exceed the length of the adjustable screw legs. Add to this the unfortunate fact that on many occasions insufficient screw legs are requisitioned and the scaffolder, or, as is more likely because its an 'easy' job, the bricklayers' labourer, resorts to packing the standards up with bricks, blocks and anything to hand. The resulting balancing act might look good in a circus ring but produces a most unstable scaffold. Standards must not be packed up in this way, regulations say so and commonsense says so, and they must be grounded on a solid even base.

Bracing

Most of the frame scaffolding systems do provide bracing members in their scheme but the supervisor must make sure that they are fitted, otherwise the erector may take the view that they are unnecessary and leave them out. This is particularly important in the case of mobile towers constructed with proprietary frames. Do make sure that all the pieces are used.

Ties

Many types do not provide tie-tubes as part of the system but this does *not* mean that ties into the structure are not required. Ties to the same formula for traditional tube and coupler scaffolding must be fitted.

Ladder access

Ladders tend to be taken for granted and are regularly abused and damaged and can then cause accidents. Ladder care should extend to proper storage, regular inspection, repair if necessary and possible, and finally to disposal of any not up to standard. 'Taking out of service' an unsafe, irreparable ladder is not enough – make sure that it is destroyed or someone will unwittingly retrieve it.

When in use, ladders must be firmly footed; lashed or clamped at the top, by lashing the stiles **not** the rungs; set at about the right angle of 75° to the horizontal and long enough to extend at least 3 ft 6 in. (1.066 m) above the point of landing.

Landing places must be provided every 30 ft (9.14 m), complete with guard rails and toeboards and care taken to ensure that openings in the platform, through which the ladders pass, are kept to a minimum size.

Maintaining the scaffold

Most scaffolds start life in good condition but many are so ill-treated and cannibalized in use that they never see the dismantling gang – they fall down first!

Larger contracts will either have resident scaffolders or specialist companies on contract to maintain scaffolds regularly and alter as required, but this is not necessarily so on the smaller site limited to low-rise scaffolding. It is hard to point the finger of blame but there is no doubt that the finishing trades, unless properly supervized, do remove ties, braces and guard rails and do not replace them.

The groundworkers are apt to undermine scaffold standards if asked to work under or near the scaffold base. Removal of the scaffold ties etc. may be absolutely necessary for, say, the glazier, but other arrangements must be made to stabilize the scaffold.

Constant supervision and a good working understanding with the trades foremen and sub-contractors is essential.

Inspection of scaffolds and register entries

All types of scaffolds must:

1. Be inspected once a week.
2. Be inspected after rough or cold weather which might have affected their stability and safety.

In the case of scaffolds from which a person could fall 6 ft 6 in. (1.981 m) or more, a register entry is required concerning the inspection and this must be made in the appropriate register, Form

91, Part 1, Section A, and signed by the person making the inspection.

Competency to inspect is important and supervisors are strongly advised to obtain instruction and training in scaffold inspection before undertaking this duty themselves. A short *aide-mémoire* follows but this is not – repeat not – a substitute for experience.

Safety check-list for scaffolds

Base – check for firm footing, adequate spread of load, no 'packed up' standards.
Check geometry – standards vertical, ledgers and transoms horizontal.
Spacing of standards – suitability for loading envisaged.
Staggering of joints in ledgers and standards.
Spacing of transoms – at least three supports for each board, not more than 4 ft (1.2 m) apart.
Guard rails and *toeboards provided.*
All necessary bracing supplied.
Means of access – ladders meeting all requirements.
Provision of ties – suitable number and position.
Proper fittings in use.
Security of any materials stacked on platforms – need for brick guards.
Overloading – dangers of shock loading when loading out scaffold with crane or fork lift.

Further reading

The Construction (Working Places) Regulations 1966. British Standard Code of Practice CP 97: *Metal Scaffolding*.

8

Lifting appliances and lifting gear

Lifting operations using machinery and lifting gear is an area of site work where risks are higher than in any other work situations and this fact led to the setting up of the Construction (Lifting Operations) Regulations in 1961. This code, as the Construction (Working Places) Regulations referred to in Chapter 7, is a long and detailed code but one that the supervisor needs to study if he is to avoid contravening the law.

I shall deal with just some of the main site operations involving lifting appliances and gear and will try to point out some of the hazards that can arise and, by the same token, be avoided.

Before doing that we should understand that the term 'lifting appliance' is defined in the regulations under Section 4 and includes some items of plant that not everyone would consider to be lifting appliances. For example, whilst most people would accept without thought that cranes or sets of sheerlegs or gin wheels were lifting appliances, many would not be so sure in the case of excavators, draglines or winches. However all these items, and others, are included in the definition and it is important that this point is remembered when the question of weekly inspections of lifting appliances is considered.

Crane operations

What should the supervisor consider when planning to bring a crane onto his site? He should first fully understand that *he* is responsible for ensuring that any hired crane, or crane owned by his own company, complies with all the statutory requirements before being taken into use on site. Not the hire company, not the plant manager but the poor supervisor bears this responsibility. Having understood this, obviously the first action he must take when ordering the crane is to require production of a test certificate, which will specify the safe working load etc. and the certificate of thorough examination of the crane. The test certificate will be issued when the crane is first put into use and needs renewal, after retesting, every four years, or following repairs or substantial alteration, and a thorough examination and recertification is required every fourteen months.

Do not let a crane onto your site until you have inspected these two certificates, checked that they are still valid and that the crane jib length specification has not been altered.

Choice of crane

There are many different types and classes of cranes available and some care is necessary to make sure that the safest yet most economical machine is chosen. The main points that should be considered are:

1. The weights and dimensions of the loads to be handled.
2. The height of lift and the distance loads must be moved.
3. The site conditions, including access for the crane, ground conditions in operating position, any special limitations that may exist, e.g. overhead lines.
4. Length of time for which the crane will be required.

Having taken all these points into consideration a machine should be selected that has a good working margin relative to both the load, taking into account the radius of the jib, and the maximum hook height. It is very important that adequate crane capacity be provided, as overreaching in turn causes overloading and leads to many collapses and cases of cranes overturning.

overreaching → overloading → overturning.

Siting of cranes

Consideration of the site conditions and the proposed operating position of the crane will have been made when selecting which crane to use, but there are two important aspects which make some amplification desirable; the suitability of the ground or area from which the crane will operate and the presence of any hazards in the proximity of the working area.

Static tower cranes – siting

It is essential that the ground on which a crane stands has adequate bearing capability. Deciding this may well be outside the capacity of the site supervisor and the advice of an experienced engineer should be sought. It is nevertheless as well to appreciate that the imposed loading will usually be a combination of:

1. The dead weight of the crane, including counterweight and any ballasting.
2. The dead weight of the load, not forgetting any lifting attachments and gear.
3. Those factors in the field of the engineer; dynamic forces caused by the movement of the crane and imposed wind forces.

Special points to be wary of are: positioning the crane close to excavations, embankments, buried mains or cellars and the possibility of storm water undermining the foundations.

Working area for mobile cranes

The same general criteria exist for mobile cranes except that their very mobility can create greater and additional hazards. The additional hazard areas can be summarized thus:

1. Travelling the crane or working it on soft ground; a particular point to consider with wheeled vehicles is the need to distribute the load from the wheels or outriggers over as large an area as possible, making use of steel or timber packing.
2. Travelling or operating on sloping ground; the movement of the load can easily affect the stability of the crane.
3. Tyre pressures on wheeled vehicles; incorrect and uneven tyre pressures can give the effect of operating on sloping ground with the same results.

Hazards in the area of crane operation

Undoubtedly the most important proximity hazard which can affect crane operations is the overhead electric line. Many fatal accidents have occurred when some part of a crane has come into contact, or even passed close to overhead lines *without actually touching*. The very high voltages carried in the lines are sufficient to allow a 'jump-over' from the lines to a crane jib and the distance that can be jumped is considerable. The voltage carried by the lines and weather conditions obviously play a part in this but jumps of up to 3 m are not uncommon with the very high voltages. This distance should *not* be looked upon as a safe working margin however and advice should be sought from the District Engineer of the Electricity Board before work in the vicinity of overhead lines is permitted.

Safe working operations

Having chosen the right crane for the job in hand and provided a safe place from which it can work the next points that the supervisor must consider are the precautions to be taken during work operations.

1. Signals to the driver should only be given by one man and he should use the recognized signals, as illustrated in Fig. 8.1.
2. Careful calculations should be made of the weights of loads to be hoisted to ensure that they are within the safe working load capability of the crane. Do *not* forget to add in the weight of the lifting gear, which may be considerable in the case of large concrete skips. Drivers and banksmen should also be reminded to watch for the possibility of loads of supposedly known weight becoming surcharged by water content, and for the suction effect of mud.
3. Check that the safety equipment on the crane, radius indicator and safe load indicator is in good working order and properly adjusted. It should be remembered, and it is sensible to remind the crane driver, that the primary function of the safe load indicator is to safeguard the crane and not to weigh the load.
4. Institute methods of securing the load to prevent it from swinging. Severe stressing of the jib can take place if loads are forced, or swing, out of perpendicular (see Fig. 8.2).
5. Never permit a load to be suspended without the operator being at the controls.
6. Make sure that sufficient clearance is maintained between the crane and any adjacent structure or materials to avoid the possibility of trapping any person when the crane travels or slews. A clearance of at least 2 ft (600 mm) is necessary.
7. Make sure that the driver is capable of carrying out the weekly inspection of his machine, that he does so and that

Fig. 8.1 Crane signals. The series of crane signals recommended by the National Federation of Building Trades Employers and Federation of Civil Engineering Contractors. The signaller should stand in a secure position where he can see the load and can be clearly seen by the crane driver. If at all possible he should face the driver. Each signal should be precise.

he makes the appropriate register entry in Form 91, Part 1, Section C.

Carrying of persons by crane

Some aspiring acrobats like to travel with a load being lifted by crane. Supervisors should be on the look-out for such lunacy, for apart from being a contravention of regulations it is extremely dangerous. The carrying of persons is permitted under certain circumstances but those carried must be housed in a properly designed chair, skip or cradle. Skips which can tip should not be used.

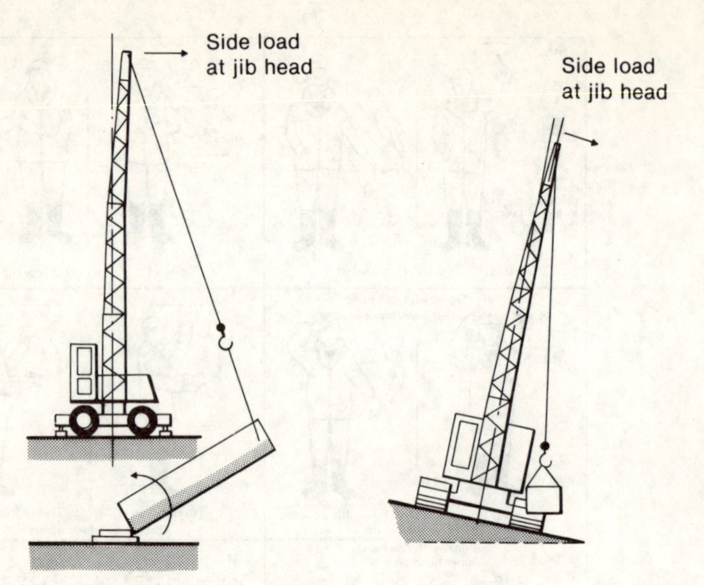

Fig. 8.2 Side stressing of jib.

Use of excavators as cranes

Excavators are designed for digging and not primarily for use as a crane, although, as has already been said, they *are* lifting appliances within the meaning of the regulations. Accordingly care must be taken when adapting an excavator for use as a crane.

Simply attaching a hook to the bucket of an excavator and lifting a load suspended on the hook is, in effect, adapting that excavator as a crane and to carry out such an adaptation without first meeting other requirements is contrary to regulations.

On the face of it the regulations do prohibit such adaptations by requiring cranes to be fitted with radius indicators and, when constructed to carry more than 1 ton, safe load indicators also. As the fitting of this apparatus to excavators is virtually impossible that would seem to be that.

The Health and Safety Executive, however, recognize that there are occasions when the temporary adaptation of an excavator is sensible and accordingly have issued an exemption certificate, providing certain criteria are met.

The adapted machine may only be used as a crane for work immediately connected with an excavation it has dug. The excavator bucket must not be removed, except in the case of a dragline, and the hook or other lifting gear must be attached to the bucket safely.

Wrapping a set of chains around the excavator bucket arms is *not* permitted. A competent person must specify the maximum load to be lifted, which must be the same whatever the radius of the jib, and which must not exceed the weight of the load which the machine in its least stable configuration is designed to lift. This maximum load must be clearly marked on the machine and also entred on a certificate, signed by the competent person, a copy of which must be kept on site.

This may seem rather a complicated procedure to have to meet but it is the law, it does give a reasonably safe working method and it can be much more economical than hiring a crane as well as an excavator.

Goods hoists

There are two main types of goods hoist, the centre slung hoist and the mobile mast hoist. The centre slung variety is built into its own scaffold tower with guides for the platform being provided on either side of the tower, whereas the mobile variety arrives on wheels complete with a small section mast up which the platform travels.

This mobility can lead to problems. Overkeen operatives can put a mobile hoist into use before it has been properly erected, unless the site supervisor is alert. The mast of this type of hoist must be tied into the scaffolding or structure at intervals of approximately 6 ft (1.82 m) before being used and, although it does not need a tower for stability, it does still require the gates and guarding which are part and parcel of the hoist tower construction (see Fig. 8.3).

Both types have similar guarding requirements and these can be summarized as follows:

1. Hoists must be enclosed at ground level by substantial enclosures and gates at least 6 ft 6 in. (2 m) high and the enclosures should extend to accommodate the engine or motor.
2. Gates of the same height of 6 ft 6 in. (2 m) must be provided at all of the landing stages.
3. It is strongly recommended that the complete hoistway throughout its height be enclosed with wire mesh so that if any material is accidentally dislodged from the hoist platform it will be contained.

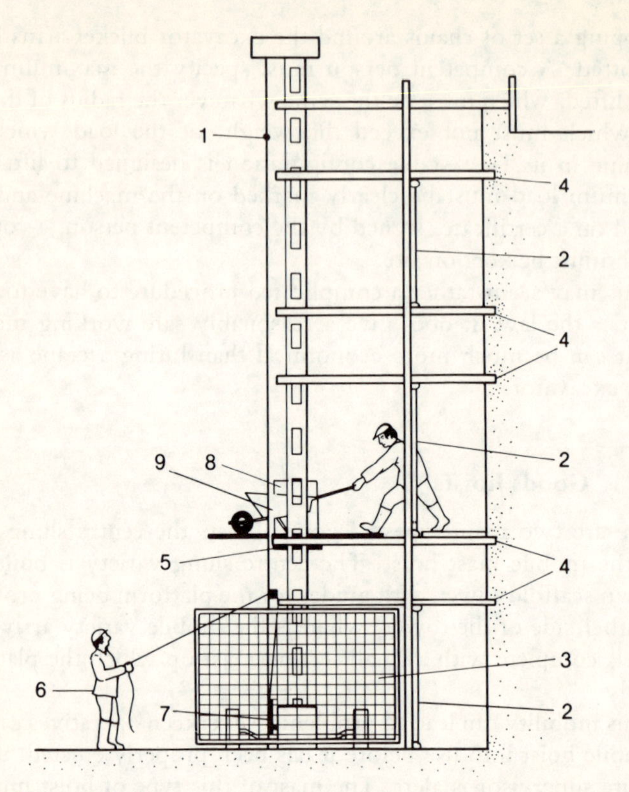

Fig. 8.3 Goods hoist.
Key: 1 Over-run device
2 6 ft 6 in. (2 m) high landing gates
3 Hoist enclosure 6 ft 6 in. (2 m) minimum height
4 Hoist mast tied into building
5 Hoist arrestor device
6 Hoist operated from one position only, giving driver unobstructed view
7 dead man handle
8 SWL marked on hoist platform
9 'Riding prohibited' notice.

Other requirements for the supervisor to check are:

1. There must be only one operating position for the hoist and the hoist driver must be trained in the job, able to see the platform of the hoist throughout its travel and over 18 years of age.

2. All materials carried on the platform must be so placed as not to be dislodged and any movable equipment, wheelbarrows, etc., must be scotched.
3. No persons shall ride on the platform of a goods hoist and a notice to this effect must be exhibited on the hoist so that it can be seen at all levels.
4. The safe working load must be plainly marked on the hoist and never exceeded.
5. Gates must be kept closed at all landing stages when not being used for access.
6. Every hoist must be fitted with an automatic device to ensure that it cannot overrun and also a device which will support the platform in the event of any failure of the ropes or gear.
7. Every hoist must be inspected once a week and a written report of the inspection made in the weekly inspection register Form 91, Part 1, Section C.
8. A further thorough examination of the hoist must be made by a competent engineer/surveyor every six months and the supervisor is advised to require a copy of this certificate when accepting a hoist onto his contract site.

Lifting Gear

'Lifting gear' is taken to mean anything used to connect the hook of a crane, or other lifting appliance, to the load to be lifted, i.e. a sling of one sort or another, and slings can be made of chains, wire rope, fibre rope or flat belting of various types.

They all have several things in common. They need to be tested before being taken into use and also after repair; they must be marked with an identifying number and given a safe working load and they must be examined every six months and the examination recorded in Form 91, Part II, Section K. This last requirement officially does not apply to items not in regular use but in practice it is safer on construction sites to include *all* lifting gear in the inspection schedule.

It should be stressed, however, that those statutory inspections are minimum requirements and the wise supervisor will make sure that more regular inspections are carried out. A chain sling can be misused and damaged the day after, or even an hour after, the six-monthly inspection, so a constant watch is essential.

Defects

These extra inspections should look for evidence of stretching of chain and wire rope slings, indicated by distorted links, rings or hooks and unequal length of multi-leg slings; cuts and nicks in the surface caused by bending the slings round sharp edged objects; wear on the outside edges of links caused by dragging chains across hard surfaces; flattening of wire rope slings and the incorrect use of bulldog clips (see Fig. 8.4).

Hooks

Special mention should be made of the hooks that are used on slings and cranes. The Construction (Lifting Operations) Regulations 1961 require all hooks either to have a safety catch to prevent the displacement of the load or to be of such a shape as to avoid the risk of displacement (see Fig. 8.5).

Prior to 1961 this regulation did not apply and of course still only applies to construction work, so there are still many hooks in use in 'factory' situations that would not be legal on site.

You may find your depot using such equipment in their work area – make sure it does not get onto your site.

Use of slings

Most people appreciate that, when using multi-leg slings, as the

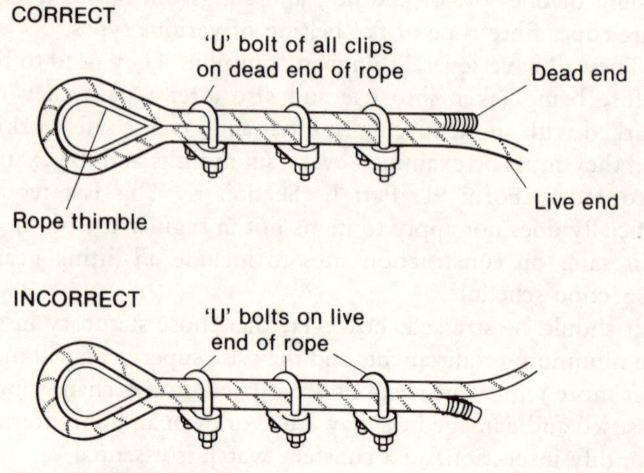

Fig. 8.4 Use of bulldog clips.

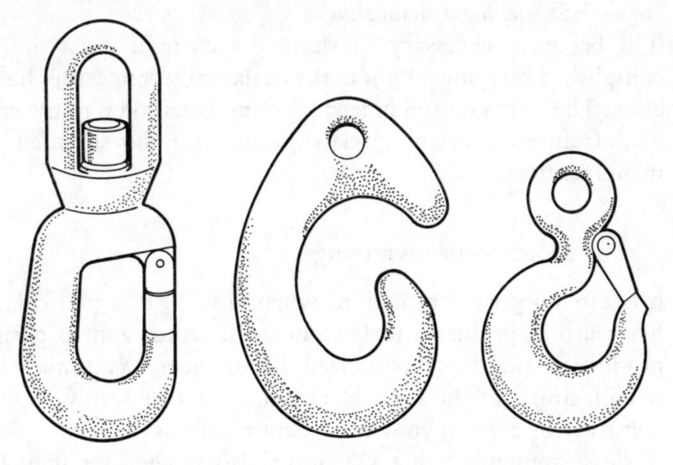

Fig. 8.5 Safe hook design.

angle between the legs is increased, so the load the sling can safely lift is decreased, but the extent of the increase of tension or load is not so generally realized. Figure 8.6 illustrates the dramatic increase in tension sustained by the sling legs when the angle is increased and highlights the need for always using slings of a length adequate to enable an angle of less than 90° to be maintained.

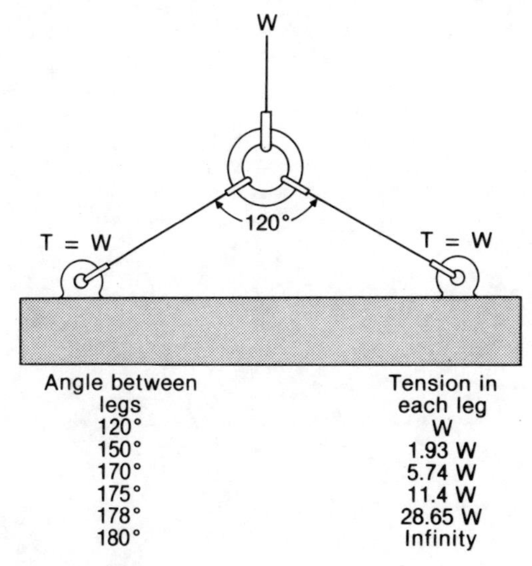

Angle between legs	Tension in each leg
120°	W
150°	1.93 W
170°	5.74 W
175°	11.4 W
178°	28.65 W
180°	Infinity

Fig. 8.6 Use of slings.

Shortening a chain sling

If it becomes necessary to shorten a chain it must not be accomplished by tying a knot in the chain or by connecting links with bolts. The only safe and indeed permitted method is to use specially manufactured shortening claws which can be supplied by the manufacturers.

Doing your own thing

It is often a great temptation, when faced with a particularly difficult lifting problem, to become an inventor and to design and produce a piece of specialized lifting gear. You may have it manufactured at the local blacksmiths or that handy engineering company or even in your own plant repair workshop.

Please remember that you must also arrange for it to be load tested and certificated before you put it to use. However safe it looks, however strong you know it to be, you cannot afford to take chances with lifting gear – there's usually somebody waiting below who collects.

9

Temporary electrical supplies

Electrical hazards can be deadly. Our natural senses, which give us warning of other visible and audible hazards, are of no assistance in unsafe electrical conditions.

For this reason, if for no other, temporary electrical installations on contract sites must be treated seriously. This is no job for the site handyman and the aim should be to make the 'temporary' installation as near as possible comparable to the standard required for permanent work.

Hazards

The hazards of poor electrical installations can broadly be put under three headings.

Electric shock

About 4 per cent of all electrical accidents are fatal and many more result in serious injuries from varying degrees of burns. Electric shock is the effect on the nervous system of the body of an electrical current passing through it. The size of the shock depends on the strength of the current, or the amperage, but as the term 'voltage' is more likely to be known and understood, it is probably easier to explain the dangers by reference to that component of Ohm's law. The strength of the shock felt increases rapidly with voltage, and a level of about 20–25 V is sufficient to cause severe pain to a person firmly gripping a live wire. It is also at about this voltage that some people find difficulty in releasing their grip as the shock may cause the muscles of the hand to contract. The higher the voltage rises the firmer the grip becomes and if the voltage is sufficiently high the current can stop the breathing, or set the heart beating very irregularly, which if it continues for more than a few minutes, can be fatal.

The exact voltages which produce certain conditions are difficult to specify as the resistance of different persons' bodies varies greatly, and the environmental conditions may also affect the issue, but it has been established that voltages below 60–70 V are reasonably safe and that the danger from voltages below 120 V is small. The national mains voltage of 240 V is, however, far from safe, should be treated with proper respect and reduced whenever possible.

Fire

Fires caused by electricity are usually the result of overloading a particular circuit with a subsequent overheating of the cables, a danger which highlights the importance of fuses in the circuit.

Also the fuses must be of the correct resistance, otherwise overloading can easily occur and you should make sure that all cases of repeated rupturing of fuses are investigated and corrected. The practice of hopefully replacing a blown fuse with a larger one is *not* a remedy you should allow on your site.

Glare

Lighting of the work area, both area floodlighting and localized supplementary lighting, is often necessary but care must be taken in positioning the lamps to avoid the danger of glare and dazzle. Localized lighting positions tend to be regularly changing and you should see that lights are repositioned above working level and mounted as high as possible. Light thrown upwards has a particularly bad glare effect and in the multi-level work situation lights serving workers at lower levels can create problems at higher levels unless sensibly positioned.

Keep a watchful eye also for carelessly placed floodlights, especially in built-up areas near busy roads, as dazzle can create hazards for passing drivers.

Site requirements

Electricity supply requirements will need to be considered as part of the preplanning exercise and the local electricity board will require to know your maximum demand during construction operations. You should compute this reasonably accurately and pass the information to the board, giving them as much notice as you can. It may be that the electricity board will wish to install the permanent supply for the building or structure straight away, in which case there would generally be an adequate supply provided, but this cannot be relied on and you must make adequate provisions.

The smaller sites will need power for lighting and heating of site huts, heating of drying rooms, cooking facilities in messrooms and canteens and water heaters for washing facilities.

A supply may also be needed for site lighting, localized and flood; lighting for hoardings; and a whole range of portable tools ranging from the drills to the wall chasers.

The larger sites may also need a 415 V supply for heavier items of plant and the grand total requirement can reach surprising dimensions; it is not an item to be guessed at.

Site Distribution

The mechanics of distributing electricity around the site will of necessity vary and will depend on the size and type of site but for the larger site, requiring a 415 V supply, the system of distribution units described in British Standard BS 4363 and Code of Practice CP 1017 is to be preferred. The system is built up from standardized components comprising a main distribution unit, which may also be combined with the incoming supply meter box; transformer units for 415–110 V 3-phase and 240–110 V single-phase supplies; outlet units, which are connected by flexible lead to the transformer units; and further extension outlets. In total it provides a comprehensive and safe system and, being comprised of units, can be used as a whole or as a semi-system thus enabling the smaller site (that may only be concerned with distributing power for a few hand tools from a supply provided to the site office) to make use of the component parts as necessary.

Cables

The necessary regular changing of position of distribution equipment and hand tools etc. makes it essential that all cables used for connecting the units and tools are flexible in nature. Flat PVC covered cable, which is used for house wiring, is *not* suitable for connecting movable equipment. It is designed for taking up a permanent position, has no flexibility and in fact regular movement can cause a fracture of the wires within the insulation. If a fracture does occur in the earth wire, the equipment may continue working but will not be protected – a very dangerous potential risk situation. The flexible cables used to connect distribution units should also be armoured as an additional safeguard and given additional protection if it is necessary for them to lie on the ground. The use of reinforced hose as a sleeve for the cable can be a helpful method. One might assume that if a competent firm of electricians are installing your site supply all cables used would be suitable. This would, however, be a dangerous assumption to make as all companies are only as good as the man they send to do the job. He may be a very good house wiring man, but site requirements are much different. Do make sure that your electrical contractor knows that the rough conditions on site and the general flexibility required make additional precautions necessary. Do not assume the electrician will know these requirements.

Voltage reduction

It has already been pointed out that voltages below 120 V give a relatively small risk of electrocution, so it must be our policy to reduce applied voltages below that level whenever possible. The large site with major items of plant operating by electricity will of necessity need to maintain higher voltages, but these pieces of plant are usually static and in consequence their supply can be installed to a permanent standard. The items that are most at risk are the portable hand-held tools and with these it is vital that, on even the smallest site, a transformer is incorporated into the system so as to reduce the voltage from mains to 110 V a.c.

Furthermore the transformer will normally be so constructed that it has a secondary winding which is centre tapped to earth. The result of this rather technical language is that in fact, although the supply will operate equipment at 110 V, in the event of a fault developing the maximum voltage to earth is 55 V, well below the danger level.

Supervisors should remember to make it clear to any subcontractors that only 110 V equipment will be provided for and permitted on site.

In some particularly dangerous work situations, however, it is desirable to reduce the voltage still further. In deep excavations and headings or when working within boilers or tanks any lighting provided should be at 50 V and again, taking advantage of a centre tapped transformer, a voltage to earth of only 25 V can arise.

Double insulated tools

Certified double insulated tools, which are used from a two-core cable and not earthed, are relatively safe and may be used on construction sites, but it is still recommended that only 110 V tools are used because of the risk of damage to the trailing leads of the supply cable.

Maintenance and inspection

Regular maintenance of all the elements of the supply is essential and, whilst not legally required, there is no doubt that a weekly visual inspection is desirable. The conscientious, or some may say careful, supervisor will do this himself as part of his normal walk round and diary the event and any defects discovered. Portable tools should be included in this inspection and particularly loose connections and damaged leads.

Inspections of fuses should also be made to make sure that no home-made fuses have been incorporated. Short 6 mm diameter bolts are unfortunately useful in this respect. Check that all equipment is provided with the proper plug and never permit adaptors as this practice will inevitably lead to overloading. Watch for the plugless cable jammed into a socket with matchsticks. One culprit, taken to task by a visiting factory inspector for such a trick, excused himself by explaining that he *had* used safety matches and not Swan Vestas! Finally check for unauthorized connections and equipment. Some very ingenious, and at the same time lethal, arrangements have been invented by tea and coffee addicts for producing that extra unauthorized cup. They have also produced fatalities.

10

Abrasive wheels

The use of abrasive wheels is controlled by the Abrasive Wheels Regulations 1970 which apply fully to any construction site. It should be pointed out straight away that the term 'abrasive wheel' includes the 'cutting off disc' used extensively in the construction industry and that all portable hand-held equipment is also included.

Regulations

The code of regulations is not very long and it is appropriate that a simplified description of the most important regulations is given:

1. All wheels shall be properly mounted.
2. No person shall mount any abrasive wheel unless:
   (a) he has been trained in accordance with the regulations;
   (b) he is competent to carry out that duty;
   (c) he has been so appointed and such appointment has been duly registered, signed and dated in a register kept for that purpose;
   (d) every person appointed shall be provided with a copy of the entry certificate of permission.
3. All practical steps shall be taken to ensure a minimum risk of injury.
4. No wheel should be fitted to a machine unless that machine is equipped with efficient starting and stopping devices that are readily accessible to the operators.
5. Where there is a rest for supporting work pieces it must be of substantial construction, properly adjusted, secured, and maintained.
6. Approved cautionary notices as to dangers shall be displayed in every area used for grinding.
7. The floor surrounding every fixed machine or area where

portable tools are being used shall be maintained in good and even condition. So far as practicable it must be kept clear of loose material and prevented from becoming slippery.

8. No employed personnel shall wilfully misuse or remove any guard, protection flanges, rest, or appliance stipulated by the regulations. Any discovered defects should be reported immediately.
9. Effectively no power machines shall be sold or hired unless statutory regulations on markings, guardings and mountings are met.
10. Where guards are specified by the regulations they will:
    (a) as far as is reasonably practicable, be constructed so as to contain every part of the wheel;
    (b) be properly maintained and secured.
11. All abrasive wheels over 55 mm diameter must be marked with permitted revolutions per minute.
12. Securely fixed to every power driven machine must be:
    (a) maximum permitted working speed;
    (b) each speed for multi-speed machines;
    (c) maximum and minimum speeds for infinitely variable machines.

I would re-emphasize the requirement to train the persons who mount or change the wheels or discs. This is the keystone of the regulations which take the viewpoint that if wheels are properly mounted then a high proportion of the hazards are removed. This code in fact does not lay down training requirements for *users* of abrasive wheels but you will appreciate that this is an aspect of training which is covered by Section 2 of the Health and Safety at Work etc. Act.

Abrasive wheel speeds

It must be remembered that centrifugal force increases not directly with speed, but at the square of the speed. The speed at which the grinding wheel revolves is, therefore, extremely important. It cannot be too strongly impressed that doubling the number of revolutions per minute of a wheel increases fourfold its tendency to burst. The peripheral speed is generally used for describing permissible wheel speeds and the maximum speeds specified by the manufacturer must never be exceeded.

On the other hand the supervisor should be on the lookout for operatives attempting to use very small cutting off discs, i.e. discs that have been so worn and reduced in size as to be ineffective.

They are ineffective, of course, because their smaller peripheral size has drastically reduced their peripheral speed and in consequence their cutting ability. The operating danger is that the operative will attempt to apply too much pressure and cause the disc to fracture.

The principal danger is that incorrectly selected, mounted or used wheels, or wheels used rotating at too high a speed, will fly apart under centrifugal force and spray the operator and anyone within range with bits travelling at the speed of bullets.

Abrasive wheel characteristics

In the process of manufacture of abrasive wheels, the abrasive and bonding materials are controlled to produce wheels of the varying qualities required for an almost unlimited range of grinding conditions and requirements.

The following are the variable elements in abrasive wheel manufacture and the standard symbols that are used to designate them:

Abrasive means the abrasive used in the wheel construction. Aluminium oxide is expressed as A, silicon carbide as C.

Grain size means the size of abrasive grains used as cutting particles. The grains are classified according to the sieve through which they have passed. The range is expressed by numbers (coarse 8 to very fine 600).

Wheel grade is generally considered as the tenacity with which the bonding material holds the abrasive grains in a wheel. Wheels are graded as 'soft' or 'hard' according to this degree of tenacity. The grade scale is expressed in letters, from A (soft) to Z (hard).

Structure means the relationship of abrasive grain to bonding material and the relationship of both to the spaces or voids that separate them. The voids or spaces in the structure assist in rapidly removing 'chips' from the wheel face thus eliminating 'loading' or choking of the abrasive surface.

Bond type means the bonding material used in the wheel construction and is described by letters V (vitrified), B (resinoid), etc.

Abrasive wheel marking system

British Standard 4481: Part 1: 1969 is now generally adopted as a basis for the marking of abrasive wheels. This specification secures uniformity and completely identifies and describes a wheel. It also provides a general indication of the hardness and grain size of any one wheel as compared with another. In view of the wide variation in grinding conditions, however, wheels of similar marking made by different manufacturers may not necessarily give the same grinding action.

The four principal wheel characteristics are marked in the following order and denoted by the appropriate symbol.

1. Abrasive
2. Grain (size of abrasive particles)
3. Grade of hardness
4. Bond type.

The system is flexible and may in addition include the manufacturer's special symbol for exact abrasive type or, in the final mark, the manufacturer's symbol *letter, numerals or both* for any special identification of the wheel type. A structure symbol may also be used between the grade and bond type markings. The marking symbols are generally shown on the abrasive wheels on a tag attached to the wheel or on an accompanying label. Tags and labels should be carefully preserved, as they give essential information for the exact duplication of a wheel.

It is essential that operators and those responsible for wheel mounting should be able to recognize the specification marked on wheels. The chart that follows (Fig. 10.1), reproduced from BS 4481: Part 1: 1969, shows the sequence of symbols used for this standard system.

Matching wheel specification to material

Selecting the correct wheel for the job is extremely important for efficient production and safety. The grinding wheel of a certain grain and grade, although a perfectly good wheel, may be almost useless and perhaps dangerous if used for the wrong work. If ever in doubt, a grinding wheel manufacturer should be consulted. In general the information that the grinding wheel manufacturer will require is as follows:

1. Speed of machine or spindle on which the wheel is to be mounted.
2. The type of grinding machine.

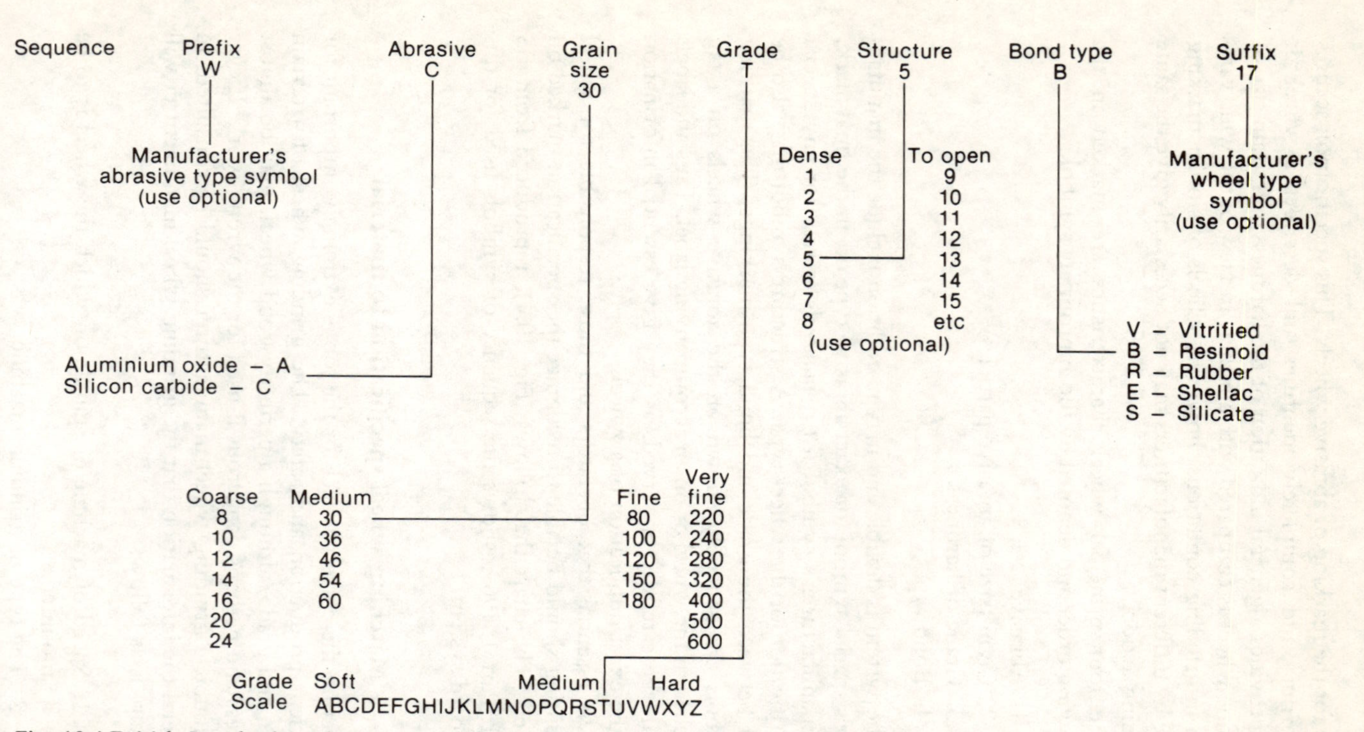

Fig. 10.1 British Standard marking system for abrasive wheels. Examples of the British Standard mark which might be found on a cutting-off wheel intended for cutting brick; the mark might be C30T5B. The prefix and the suffix of the marking system are manufacturer's options and usually omitted.

3. The material to be ground or cut.
4. The rate of stock removal required.
5. Accuracy and finish required.
6. The area of wheel contact, i.e. wheel diameter required.

The first factor to be considered is the material to be ground or cut. For materials of high tensile strength select a wheel having aluminium-oxide abrasive content. For materials of low tensile strength select a wheel having silicon-carbide abrasive content.

Generally speaking hard materials require a wheel having fine grain with close grain spacing whilst soft ductile materials demand a wheel having coarse grain and wide grain spacing.

The amount of material to be removed and the accuracy of finish required demand different characteristics from the grinding wheel. For fast cutting and rapid stock removal a coarse grain and wide grained spacing is required. For a fine accurate finish the wheel having fine grain and close grain spacing is demanded.

The third factor affecting the choice of a wheel is the area of contact between the wheel and the work. Generally this is judged by the wheel diameter. A large area of contact or large wheels generally demand coarser grain of soft grade with wide grain spacing. Wheels having a small area of contact demand a harder grain with closer grain spacing.

The type of grinding machines being used often influences very much the selection of an appropriate wheel. Heavy, rigidly constructed machines will take softer wheels than the lighter, more flexible types, such as hand-held portable machines. The condition of the grinding wheel machine itself can also play an important part and it is important for optimum performance that all machines be kept in top condition.

Common wheel mounting methods

Straight and tapered wheels – both large hole and small hole

The wheels should be gripped between two flanges of equal diameter, the inner flange being keyed to the spindle, the other being tightened by a nut on the threaded spindle end. Excessive tightening of the nut is unnecessary. Each flange must have equal recesses and the washers should be just larger than the flange diameters (see Fig. 10.2).

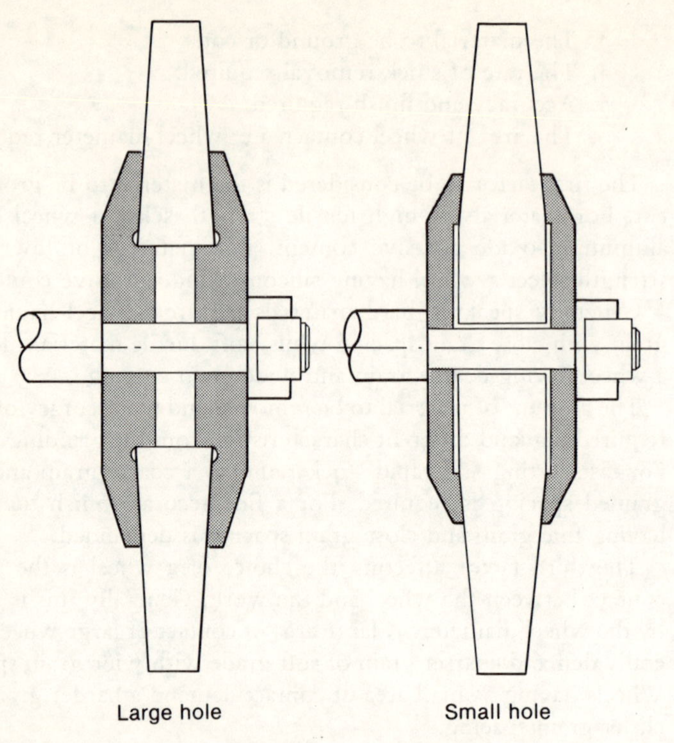

Fig. 10.2 Straight and tapered wheels.

Threaded and unthreaded hole wheels

Two principal methods are accepted for mounting cup wheels on portable grinders embodying the use of wheels with either threaded or unthreaded holes.

Threaded hole wheels. These are commonly mounted by a method involving the use of threaded steel bushings. The wheels are screwed onto the end of the threaded machine spindle, which is fitted with a flange. The flange should be flat and not recessed, thereby avoiding strain on the threaded bushing.

Unthreaded hole wheels. These are generally mounted by means of an adaptor flange or flange nut. In this instance, the backed flange and adaptor should be equally relieved to provide proper support for the wheel (see Fig. 10.3).

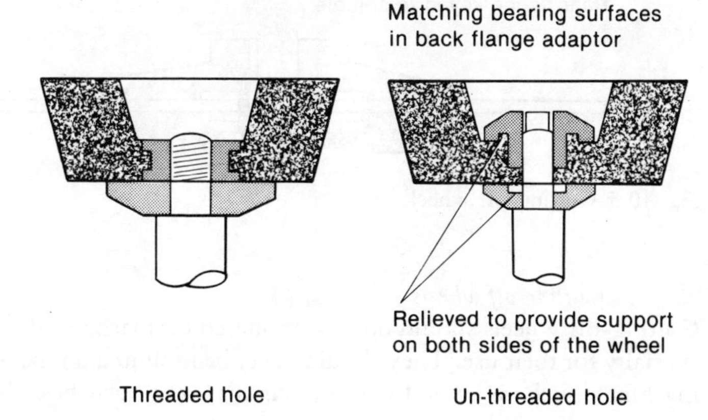

Fig. 10.3 Cup wheels.

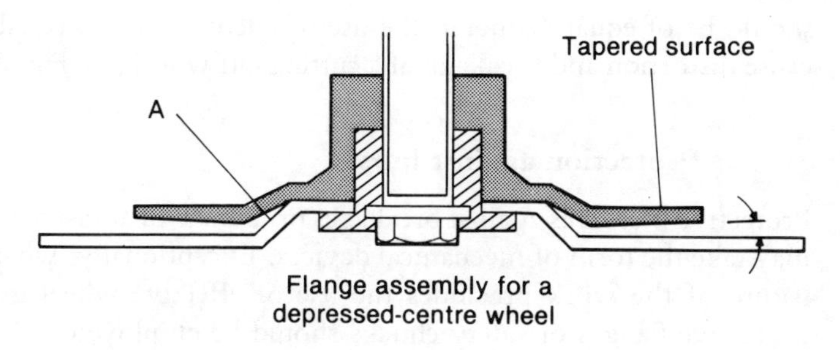

Fig. 10.4 Depressed centre wheel.

Disc wheels

Depressed – centre wheels

Depressed – centre wheels should only be mounted with a flange assembly as shown in Fig. 10.4. Two points should be noted:

1. When the adaptor has been tightened there should be a slight clearance between the flange and the wheel at (A). This ensures that clamping pressure is exerted only at the centre of the wheel.
2. The outer part of the face of the flange adjacent to the wheel should be tapered as shown. This allows the full width of the flange to support the wheel during the grinding operation.

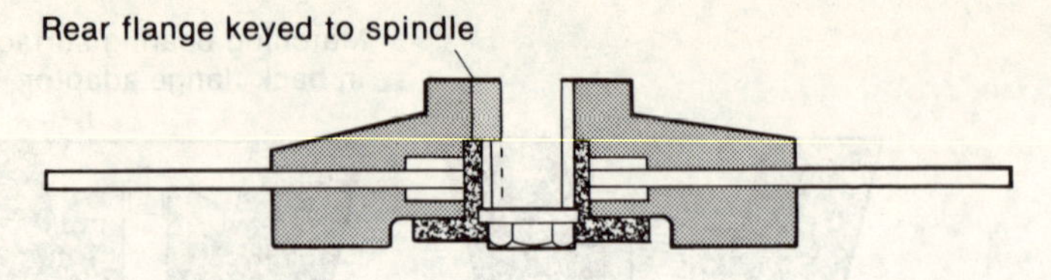

Fig. 10.5 Cutting-off wheel.

Cutting-off wheels

Cutting-off wheels should only be mounted on machines designed specially for their use. They should never be mounted on makeshift machines such as woodworking circular saw benches. Never mount an unreinforced cutting-off wheel on a portable grinding machine. The wheel must be of the reinforced type.

Flanges should be as large as practicable and never less than one-third of the wheel diameter. It is most important that the flanges should be of equal diameter; the use of unequal flanges is liable to cause distortion and breakage of a cutting-off wheel (see Fig. 10.5).

Protection against bursts

Protective measures against breakages or 'bursts' of abrasive wheels may take the form of mechanical devices. Exceptionally, where the nature of the work precludes the use of effective wheel guards, protective flanges or safety chucks should be employed.

Guards have four main functions:

1. To contain the wheel parts in the event of a burst.
2. To protect the wheel against inadvertent damage from shock, etc.
3. To prevent, as far as possible, the operator coming into contact with the wheel.
4. To prevent an oversize wheel being fitted.

Type of guard

Where the wheel is used for periphery work over a limited arc, hoodtype guards are used. The aim is to enclose the wheel to the greatest possible extent, the opening being as small as is consistent with the nature of the work.

Some examples of guards for portable machines are shown in Fig. 10.6.

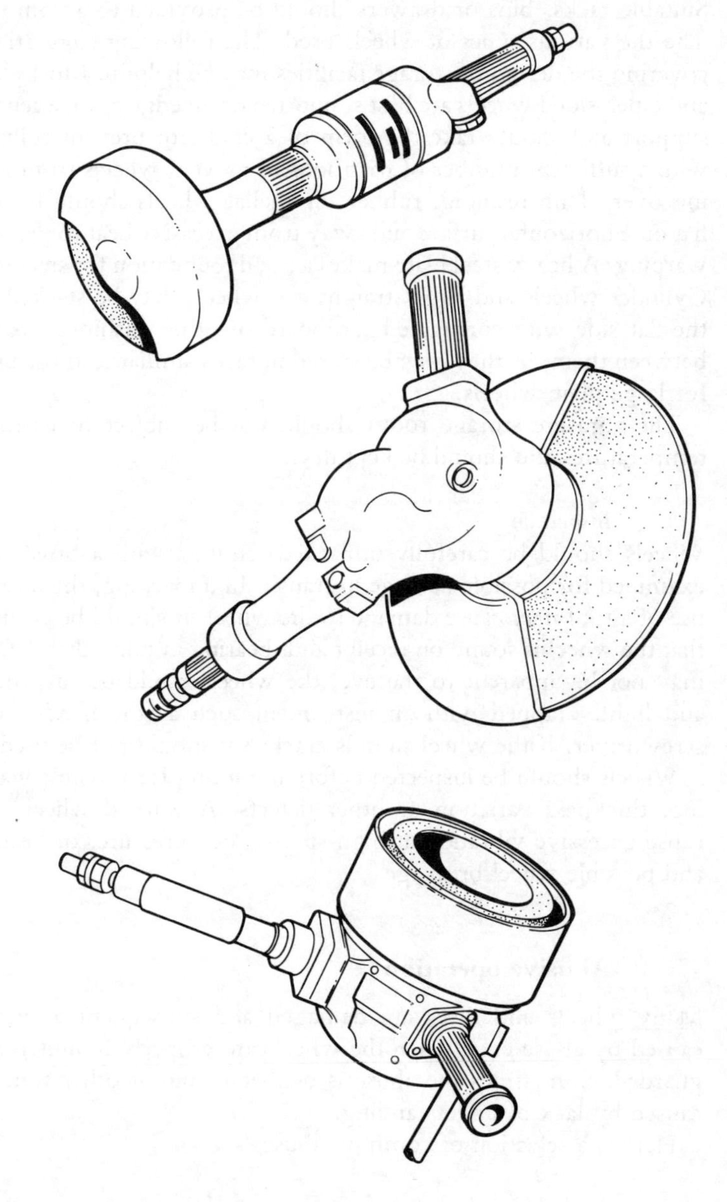

Fig. 10.6 Guards for portable machines.

Storage and inspection of wheels

Storage

Suitable racks, bins or drawers should be provided to accommodate the various types of wheels used. The following suggestions covering the design of storage facilities may be helpful. Most plain and taper-sided wheels are best supported on an edge or on a central support and should take the form of a cradle to prevent rolling, with a sufficient number of partitions to prevent wheels from falling over. Thin resinoid, rubber, or shellac wheels should be laid flat on a horizontal surface and away from excessive heat to prevent warping. A heavy steel plate makes a good foundation for stacking. Cylinder wheels and large straight cup wheels may be stacked on the flat side with corrugated cardboard or other cushioned paper between them, or they may be stored in racks similar to those used for large plain wheels.

The abrasive storage room should not be subject to extreme temperatures and should be kept dry.

Inspection

Wheels should be carefully unpacked, cleaned with a brush and examined for possible damage in transit. In unpacking, the unwise use of a tool may cause damage to the wheel. It should be ensured that the wheel is sound on receipt and, bearing in mind that defects may not be apparent to the eye, the wheel should be suspended and lightly tapped with an instrument such as the handle of a screwdriver. If the wheel sounds cracked it should not be used.

Wheels should be inspected before mounting for possible warping, thickness variation or other defects. A warped wheel may cause excessive vibration at high speeds, side pressure, or heating and possible wheel breakage.

Abusive operation

Many wheels and discs are damaged and subsequent accidents caused by abuse even when the wheels are properly mounted and guarded. Sometimes the abuse is deliberate and at other times is caused by lack of understanding.

Here is a selection of common abuses.

Lack of speed control

The lack of control or improper adjustment of machine speed, particularly with air grinders, results in considerable trouble in the field. It is not uncommon to find cases where speeds are too high even for organic bonded, high-speed wheels. In the case of air-driven grinders, this may be caused by the speed governor getting out of adjustment or not functioning at all. There are cases on record where operators have removed the governors entirely, in order to make the machine run faster. Portable tools leave the factory ready to deliver the maximum power at the correct speed, but too often these values are not checked at frequent enough intervals.

A point to remember with portable air-driven tools is that the pressure indicated on the air gauge at the compressor may be misleading. The pressure at the tool is more important and may be quite different, due to line loss. Too high a pressure is dangerous and too low a pressure causes loss of efficiency. Follow the machine tool manufacturer's recommendations for rated pressures.

In the case of electric machines, the delivered voltage should be periodically checked.

Inadequate power

Wheels can be broken by lack of adequate power. If a wheel slows down materially when being applied to the work with normal pressures, it is an indication that sufficient power is not being transmitted to the wheel spindle.

The slowing down of the wheel due to inadequate power is also likely to cause the wheel to develop 'flat spots', leading the operator to believe that the wheel has hard and soft spots. The bumping action resulting from the flat spots may ultimately cause wheel breakage. If the machine is 'underpowered' for the job, the solution is to get a higher-powered machine.

Grinding on flat side of straight wheels

Grinding on the flat side of straight wheels is hazardous and should not be allowed on such operations when the sides of the wheel are appreciably worn thereby, or when any considerable or sudden pressure is brought to bear against the sides.

'Cramping' of straight wheels

With the same thought in mind to prevent dangerous side pressures on straight wheels, care must be taken when using a portable grinding machine for cleaning out grooves or between projections. A 'cramping' action on the wheel is dangerous and a cause of breakages.

Dropping machine on floor, etc

Portable grinding machines are probably subjected to more abuse in handling than any other type of grinding equipment. It is easy to damage a wheel by careless handling of the machine. This can occur when the wheel is at rest as well as when in use. Dropping it to the floor or bench, bumping it into castings and obstructions, or setting it down carelesssly can all damage a wheel, causing the wheel to be broken when brought up to speed.

Protective equipment and procedures

Among the most common accidents to operators of grinding machines are eye injuries. The majority of eye accidents would be prevented by the provision and use of suitable goggles or effective screens.

It is essential that the appropriate eye protection for any particular work be provided and used, especially in all dry grinding and cutting off operations. The screens are particularly useful on stationary machines which are used for short periods by different workers and where the operators are often reluctant to wear goggles. In all instances, however, it is better to play safe and wear goggles as well.

Goggles should be comfortable to wear, allow a large clear field of vision, and have the best possible ventilation. The best results are obtained by experimenting with various types before a final decision is made and by individual issue to workers. Adequate supervision of issue and use is essential.

Reference should be made to the Protection of Eyes Regulations 1974 and to Chapter 25 of this book.

Notices required

When abrasive wheels are used on a contract site the display of certain placards is required by regulation. Ideally they should be posted adjacent to a fixed machine or, if only portable equipment is in use, then the site notice board is the place.

The placards required are:

Form 2345 – Abrasive Wheels: placard.
Form 2347 – Abrasive Wheels: precautionary placard.
Form 2351 – Abrasive Wheels: maximum permissible speeds.

The register of appointment of persons to mount abrasive wheels is Form No. 2346.

11

Cartridge operated tools

Cartridge operated tools provide fast and simple fastenings into concrete, steel and brickwork and, when operated by trained and experienced workmen, are no more likely to cause an accident than any other fixing method. By the same token, in untrained and inexperienced hands, the tools can be very dangerous indeed and many fatal and serious injuries can be laid at their door.

How do the tools work?

There are two main types of cartridge tools:

1. High velocity – where the exploding cartridge fires the fixing device down the barrel of the tool in free flight and so into the workpiece.
2. Low velocity – where the exploding cartridge acts on a retained piston which drives the fixing device into the work.

The low velocity tools can additionally be sub-divided as Fig. 11.1 shows.

It should be appreciated that the high velocity tool has a muzzle velocity approximately equivalent to a small firearm and there have been reported incidents when injuries have been caused from a distance approaching 500 m. The low velocity gun has roughly a quarter of the power but can still be very dangerous.

Major hazards

There are three main hazards which can arise when using cartridge operated tools:

1. Firing through a soft material,when the fixing device

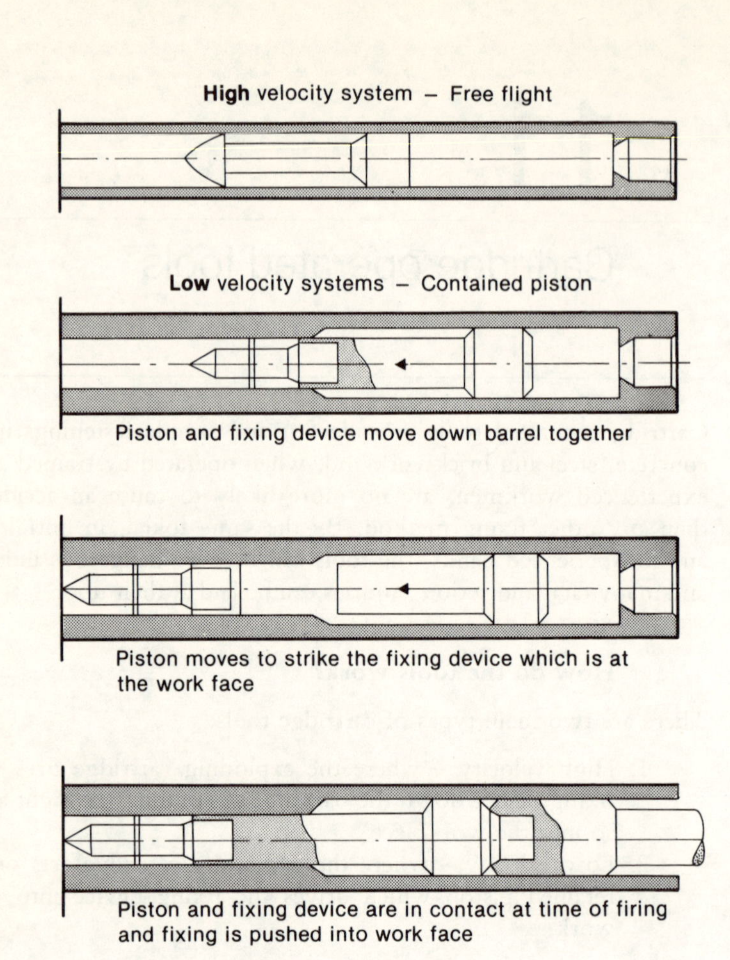

Fig. 11.1 Cartridge tool types.

punches its way right through and becomes a missile.

2. The possibility of a ricochet, when the fixing device is deflected and can turn back on itself and strike the operator or another person.
3. Splintering of the material fired into, and the consequent danger of the operator and other workers being struck by splinters.

Secondary hazards are the possibility of hearing damage, particularly in confined spaces; and physical strain, bruising or the operator being thrown off balance by the recoil action of the tools.

It should also be remembered that the cartridge, when exploding, is a violent source of ignition and consequently cartridge operated tools must never be used in work areas or situations where flammable vapours are likely to be present.

Avoiding the hazards

Cartridge strength

One of the main causes of all the hazards is the use of cartridges of a higher power than is necessary. There are, of course, several different types of tools offered by competing suppliers and all have slightly differing systems of cartridge strength identification. They do all, however, *have* a range of cartridges and you should make sure that your operators follow the golden rule of using the weakest cartridge possible.

A system of colour coding is adopted as a means of identifying the varying strengths of cartridges but the codes do not entirely match up for different makes of tool and the only safe way is to use the cartridges as specified by individual manufacturers.

Never attempt, or permit your operators, to mix cartridges or fixing devices, as they are only suited to the equipment for which they were manufactured. A final point in connection with cartridges is to check that operatives using tools do not have a colour vision defect.

Tool guards

Most guards fitted to cartridge tools are sprung and the firing mechanism so arranged that the tool will not fire unless the guard is held firmly against the workpiece. This is a good example of a hazard being 'engineered' out of the operation. The guard is an important safety feature and must not be tampered with. Special cutaway guards can be obtained for dealing with fixings in particularly difficult positions and, if these are used, extra care must be taken.

Personal protection

All guards fitted are of necessity very small and consequently provide little protection in cases of splintering of the workpiece. Goggles should always be worn as a secondary protection and are, in fact, required to be worn by the Protection of Eyes Regulations 1974.

The problem of noise has already been mentioned briefly. This may not be a problem in open situations and when only a few fixings are to be made, but in confined areas there is a real need for ear protection to be provided. Safety helmets are also a necessary precaution and should always be worn by cartridge tool operators and by other operatives in the work area.

Operating precautions

Complete penetration

If the material to be fixed to is soft enough to make firing through possible when using the weakest cartridge then obviously this type of fixing is neither suitable nor necessary and another method must be adopted. The real danger arises when a fix is generally made satisfactorily but where there is the possibility of voids or soft spots in the material. In these circumstances the operator should always check the area beyond the point of work to be make sure that no one is in the line of fire.

Ricochet

The most common causes of ricochet are;

1. Failure to hold the tool tightly at right angles to the work.
2. Attempting to fire into a hole already made by a failed fixing. If a fixing fails to hold the next attempt must be at least 50 mm away.

Splintering

While the possibility of splintering is greater with brittle and glazed materials it is still a hazard with concrete, brick and masonry, particularly when an attempt is made to fix too close to the edge of a column or brick work. The 50 mm rule should again be applied and fixings kept at least that distance from any edge.

Recoil hazards

Instruction and training can virtually remove this hazard. If the operator uses the tool sufficiently to become familiar with the possible effects of the recoil before attempting to fix from a platform of any sort, then there should be no problem.

Training

Proper training is, of course, necessary for operators, not only to prepare them for a possible recoil, but also to deal with the other more important aspects of cartridge tool use.

The manufacturers, appreciating that these inherently hazardous tools need specialized instruction, are generally prepared to offer free training in the safe use of their equipment, and these facilities should be accepted.

The instruction given will be comprehensive and will deal with cartridge strengths, recommended fixings, operating procedures, cleaning and maintaining the equipment and, by no means the least important, the procedure to be taken in the event of a misfire.

Misfires

Many accidents have occurred through operators not appreciating and carrying out the correct misfire procedure. If a misfire occurs the operator must continue to hold the tool in the firing position against the work surface, re-trigger and attempt to fire a second time. If the tool still fails to fire it should be held firmly against the work surface for at least a further fifteen seconds to guard against any possibility of a slow burning primer, and then the cartridge should be moved as recommended in the manufacturer's instructions.

Any cartridge that has failed in this way should be returned to the supplier representative, when next he calls, for destruction.

Storing and maintenance

All cartridges not issued for immediate use should be stored in a cool, dry place and in a secure locked area. Their issue should be carefully controlled and kept to the minimum necessary. At the end of the work period all cartridge tools should be cleaned and lubricated and stored under lock and key. This regular maintenance is a vital part of the safety precautions and if a tool becomes unreliable despite regular maintenance you should return it to the supplier for attention. Never attempt other than routine maintenance on site and in any case arrange for your supplier to give each tool a thorough overhaul at least once every year.

12

Compressed air tools

Compressed air tools are used on most construction sites and a wide variety of tools is available. Compressed air is a useful power source, but one that has its share of hazards.

General hazard

The hazard that is ever present with all types of tool is the compressed air itself. Serious injuries are regularly sustained by, or are caused by, operatives who are unaware of the dangers, and the supervisor must ensure that these dangers are appreciated by all concerned.

A hose must never be directed towards a person's body and the practice of workers using compressed air to dust down their clothes is extremely dangerous and must not be permitted. The greatest danger lies in air being forced into one of the body openings. Even a pressure as low as 103 kN/m^2 has been known to cause injury and it has been estimated that a pressure of only 27 kN/m^2 is sufficient to rupture the bowel. Eyes and ears are perhaps particularly vulnerable and many cases of 'foreign body in eye' and perforated ear-drum are reported annually.

More frightening, however, is the extent of the injuries that can be caused by air entering the rectum and unfortunately this is a situation which happens too frequently. Compressed air seems to have an attraction for the practical joker and one reported 'joke', when an apprentice applied an air hose to the behind of another lad, resulted in the following injuries: there was extensive bruising and bleeding in the area of the rectum; air had been forced through the tissues of the abdomen, chest and neck; the hernia channel in the groin was ballooned with air; the abdomen was filled with air; the lower bowel was torn open in three places; the abdominal cavity was filled with material from the lower bowel; the lining of the abdominal cavity was torn. Despite an immediate operation and intensive care the lad died within three days as a result of his injuries. All of these injuries were caused by an air pressure of only 241 kN/m^2.

The penetration of air through a cut or scratch can also have serious results and can cause the affected part to swell to an alarming degree and to be extremely painful. If the air then enters the bloodstream it can prove to be fatal.

I am sorry to have gone into such clinical detail but would repeat that incidents such as this are regularly reported and will continue to be so reported until everyone appreciates the dangers, or possible dangers, of this very useful form of power.

Particular hazards

More particular hazards can arise from the specialist tools that are operated.

Grinding machines

Particular attention should be paid to the dangers of overspeeding of air-operated grinding machines. All tools will be fitted with air regulators and unless these are properly maintained and lubricated in accordance with the manufacturer's instructions they will not accurately regulate the air supply and overspeeding and bursting of the wheel may result. Do not forget that the wearing of goggles is essential when operating this type of equipment.

Wood borers or augers

Used extensively in heavy civil engineering and marine work to bore holes in heavy timbers, these machines also need careful operation. The length of the drill, of necessity some 600 mm, makes guarding virtually impossible and a real danger exists of clothing becoming entangled with the moving drill bit.

It is essential that all operators are instructed to wear clothing that cannot become entangled; to hold the drill only by the handles and to be standing on a firm footing; to put the drill into motion only when it is in contact with the work and not to extract it or put it down until it has stopped revolving.

Road/concrete breakers and spades

The most regular cause of accidents with this type of equipment is a failure of, or incorrect setting of, the tool retaining clip. This permits the tool point or spade to be accidentally expelled from the body of the equipment and usually results in either someone below being struck by the falling point or spade, or the operator himself sustaining injuries from the dropping tool. Operators must play their full part in preventing this type of accident by reporting any malfunctions of the equipment but they *will* need encouragement to do this and this means that the supervisor must welcome this 'criticism' of the equipment. Many accidents have occurred as a direct result of an operator struggling on with faulty equipment because he knew any complaint would be looked upon as being 'bloody-minded'.

Improper use of points and spades can also lead to accidents. They are not designed to be used as crowbars and will easily break if stressed in this way, letting the body of the tool fall – usually onto the operator's foot. For the same reason it is important that the tools are kept sharp and properly tempered.

Operators should be trained to keep their feet out of the line of fire, to keep them on a level with or above the level of the work and, as a final line of defence, to wear steel toe-capped footwear.

The compressor plant

Most compressors used on construction sites will be of the mobile variety and the supervisor needs to be reminded of his responsibility to ensure that all plant on his site is safe to be used. In the case of compressors this extends to being satisfied that the air receiver has been thoroughly examined and certificated within the last 26 months; that all moving parts are adequately guarded; that clear operating instructions are displayed on the compressor; that the air receiver is equipped with a safety valve and a pressure gauge and that the maximum safe working pressure is clearly marked.

When operating it is important that the air supply *to* the compressor is clean and not contaminated by any flammable or explosive mixtures, or the exhaust gases in the case of any petrol or diesel engined compressor. Air filters are normally fitted and this is particularly important if the tools are to be used in confined spaces when special attention must be given to this point.

Hoses and hose connections

All hoses for portable tools must be designed for the pressure at which it is intended to use them. They should be kept as short as possible and given protection from site and general traffic. Connections must be properly clamped, as any failure of a connection when under pressure can lead to the hose whipping violently.

For the same reason no adjustments or repairs must be made until the supply has been shut off and air cleared from the system.

Noise problems

The question of noise is dealt with more fully in Chapter 29 but it would be wrong not to give the subject a mention here as undoubtedly the road breakers create many noise problems.

Much can be done in noise reduction by fitting mufflers to the body of the tools and this method should be adopted whenever possible. The nature of the operation however – breaking up concrete or asphalt surfaces – makes any noise reduction at the business end unlikely, as it is thought that rubber points or spades would be somewhat ineffective. The only action that can be taken is to provide the operator and other workers in the vicinity with ear protection.

13

Highly flammable liquids and liquefied petroleum gas

These substances are the subject of a special code of regulations, the Highly Flammable Liquids and Liquefied Petroleum Gases Regulations 1972, and this code sets out requirements for the *storage* of liquefied petroleum gases and highly flammable liquids and also regulations regarding the *use* of highly flammable liquids.

Definitions

Highly flammable liquid is defined as being a liquid, liquid solution, emulsion or suspension which gives off a flammable vapour at a temperature of less than 32 °C and includes many paints, enamels, sealers, adhesives and solvents in addition to the more obvious flammable liquids. Liquefied petroleum gas means commercial butane, commercial propane or any mixture thereof and is sold under many trade names, for example, Calor Gas, Calor Propane, Bottages, Propagas, Glogas etc.

Storage

Highly flammable liquids

The quantity of substances in this category that it is necessary to store on site will obviously vary tremendously and, whilst it may be sensible from the safety viewpoint to keep this to a minimum, storage of sizeable quantities may be unavoidable. The regulations permit small quantities to be kept in the work area or in a general store so long as the total amount does not exceed 50 litres and provided the containers are kept in a suitably placed cupboard or bin constructed of fire resisting material. The bin or cupboard must have a retention sill to contain any spillage.
Quantities above 50 litres must either be stored in a safe position

in the open air or in a storeroom which is either in a safe position or is constructed of fire-resistant materials. Again retention sills are necessary.

Every storage area, bin or cupboard must be clearly marked 'Highly Flammable'.

Liquefied petroleum gas

Cylinders of liquefied petroleum gas must be kept in safe positions in the open air or, where this is not reasonably practicable, in a storeroom constructed of fire-resistant material which is not used for any purpose other than for storing liquefied petroleum gas or acetylene cylinders. The cylinders must be stored standing upright with the valve uppermost to avoid the possibility of any leak of liquid. These storage areas and rooms must also be clearly marked 'Highly Flammable – LPG.'

Two things should be emphasized in respect of LPG storage:

1. The days of having two or three spare cylinders under the foreman's desk, or tucked away in the general store, have long gone and they should be stored in the open air. Ideally you should provide the cylinders with a security compound of heavy wire mesh in an open situation not within 1 m from other buildings or the boundary fence, or within 3 m from a store containing oxygen cylinders or highly flammable liquids.
2. The regulations say 'in the open air unless this is not reasonably practicable' and I would remind you that the term 'reasonably practicable' has been defined in the courts and requires you to weigh the possible risk against the cost involved. The risks involved with the storage of LPG are considerable and consequently a court would expect reasonable costs of providing open-air storage to be met. In any event the alternative of providing a fire resisting structure can be just as costly.

Hazards in use – highly flammable liquids

Fires and explosions caused by highly flammable liquids occur regularly on construction sites and in workshops and stores associated with sites. Paints, thinners, solvents and adhesives are to be found on most sites and relatively small quantities are sufficient to produce the risk.

Vapours

All flammable liquids produce vapours and it is these rather than the liquids themselves which are ignited and which burn. Highly flammable liquids give off these vapours at ordinary room or outdoor temperatures and unless they are kept in closed containers the vapours will mix with the atmosphere and create an ignitable and possibly an explosive mixture.

Nearly all vapours produced from flammable liquids are heavier than air and so tend to sink to ground level and act rather like a liquid, spreading out and creating a widespread hazard. This can be particularly dangerous in trenches, excavations or basements as the vapour may accumulate and can remain undispersed for a long time. It is important to appreciate the extent of the danger. An open container of highly flammable liquid can produce a hazardous atmosphere over a wide area in an unventilated situation. For example it has been calculated that one litre of acetone is capable of creating a flammable atmosphere of about 12 m^3. Or, put another way, if it was left to vaporize could cover a room of 40 m^2 with an explosive layer of acetone vapour approximately 300 mm thick.

It is true to say that pure acetone is not found every day on a construction site, but other materials *are* used that have flash points well within the acetone range and obviously can produce similar risks.

Flash points

The flash point of a liquid is the lowest temperature at which it gives off enough vapour to form an explosive mixture with air. Consequently when liquids with flash points at or below the existing temperature are exposed to the air they will immediately release flammable vapours. Liquids with higher flash points than the existing temperature only give off vapours when heated but the possibility of accidental heating of, say, white spirit should not be disregarded.

Some examples of common 'highly flammable' liquids and their flash points are: petrol −43 °C, methylated spirits 13 °C, acetone −18 °C, toluene 4 °C.

Substances which fall into the lesser risk area of 'flammable' liquids include white spirit which, with a flash point of 38 °C, is only 6 degrees from the higher classification.

Paints and adhesives can fall into either category and careful checks should be made to ensure that proper precautions are taken as necessary.

Sources of ignition

A weak spark or cigarette can be all that is needed to ignite flammable vapours, so where flammable liquids are in use great care must be taken to guard against sources of ignition.

Electrical equipment

Electric switchgear and fittings produce hot surfaces and sparking sufficient to be an ignition source. Much switchgear and most of the lighting fittings are positioned so as not to be normally affected by 'heavier-than-air' vapours, but many switched sockets are sited at or about skirting level and their use with electrical equipment has been the cause of serious accidents. Obviously in a workshop using highly flammable liquids on a regular basis proper intrinsically safe, non-sparking fittings would be necessarily provided but this is not normally the case with the more usual building under construction and accordingly is a point to be watched.

Hand tools

Ordinary metal hand tools can produce sparks when striking metal, stone or concrete and equipment of this sort must not be used in the presence of flammable vapours. Non-sparking tools made of copper alloy or aluminium bronze can be obtained but the most satisfactory safeguard is to prohibit work requiring hand tools in areas where highly flammable liquids are being used.

Human carelessness

Frequently fires and explosions are caused by human carelessness and thoughtlessness and there have been innumerable cases where a lighted match or cigarette has been the ignition source. It cannot be stressed too strongly that smoking and highly flammable liquids do not mix and the supervisor must make sure that the operatives dealing with a dangerous substance are made aware of the hazards and that a no-smoking rule is enforced.

Ventilation

Where small quantities of vapour are released out of doors and away from sources of ignition there is little danger of explosion but where the work is inside a building some form of ventilation is necessary.

The object is to dilute and remove the vapour so that the concentration is kept to a safe level and, in addition to the explosive risk, one must appreciate that some flammable vapours are also of a toxic nature and that consideration must also be given to this hazard.

The ventilation method may be either natural or induced by mechanical means. Natural ventilation, with permanent openings at both high and low levels, is usually adequate for low concentrations but it is rather unpredictable from day to day as it depends on wind speed and direction and its effectiveness should be regularly checked. If the natural air movement is found to be sluggish then mechanical assistance should be provided and care taken to ensure that any electrical motor or switching is positioned outside the work area and in a safe situation.

Hazards in use – LPG

Liquefied petroleum gas has already been defined in this chapter as being either butane or propane or a mixture of the two and, as has also been explained, is distributed under various trade names.

The main difference between the two types are:

1. Pressure in the cylinders.
2. Difference in boiling temperatures.

Butane is bottled at a pressure of approximately 172 kN/m^2 whilst Propane is at a higher pressure of approximately 620 kN/m^2. This immediately draws attention to the need for supervisors to ensure that equipment fed by the LPG supplies is correctly matched to the supplying gas. The manufacturers make an attempt at ensuring a proper match by producing butane equipment with a larger bore supply nipple than is provided for equipment designed to be fed by propane, but no one is more inventive than the construction site handyman and many cases have been reported of ill-matched equipment leading to gas leaks and subsequent fires and explosions.

The difference in boiling temperatures is also significant as, once the valve is opened and the pressure relieved, the liquid does, in fact, boil and in so doing produces the gas. Butane boils at –1 °C and propane at –42 °C and so it will be seen that the use of butane is somewhat limited in extreme weather conditions whilst propane remains usable.

LPG characteristics

Apart from the difference in pressure and boiling point the two types of liquid gas are very similar:

1. They are non-toxic and so are non-poisonous but are, however, slightly anaesthetic when high concentrations are inhaled.
2. One volume of liquid, if released, almost immediately expands to approximately 200 volumes of vapour and accordingly any leak, especially of liquid, must be treated very seriously (see Fig. 13.1).
3. The gas is twice the weight of air, volume for volume, and so will find the lowest point and may be difficult to disperse.
4. A very small proportion of gas in air can produce an explosive mixture.
5. Liquid propane, because of the very low temperature limits, can cause frost-burn if it comes into contact with the skin.

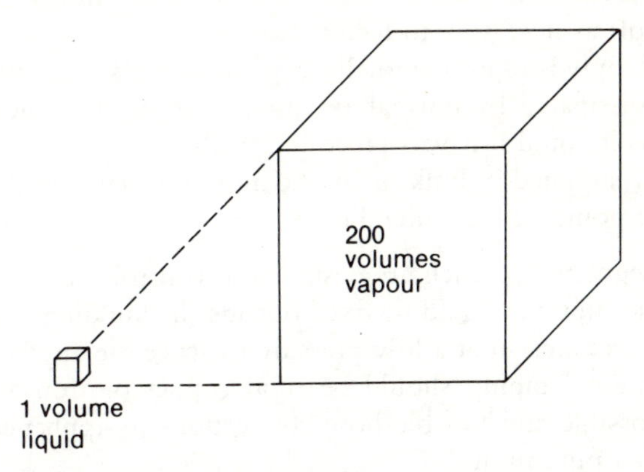

Fig. 13.1 Danger of leak of liquefied petroleum gas.

Transporting and handling

Cylinders are fairly robust in construction, and rightly so, but they still need to be carefully transported onto and around the site. Never permit cylinders to be dropped or tipped from a dumper and always arrange that they are transported whilst standing upright. Never permit the moving of cylinders by crane or other lifting appliance unless they are carried in a specially constructed container. One of many incidents reported tells of a propane cylinder which slipped from an improvised carrying cradle, fell about 9 m and struck some structural steelwork which knocked off the cylinder valve. The large volume of escaping gas was ignited by one of many possible sources, and the resulting explosion seriously injured sixteen workmen.

Particular applications

LPG is a particularly portable energy source and serves a wide variety of implements, heaters, blowtorches, boilers, floodlights, etc.

Site huts

When used for site accommodation LPG can also be supplied on site as a bulk installation and, for the longer running contract, this system can have advantages. A bulk installation tank is shown in Fig. 13.2.

The gas is piped from the supply tank to the office block and ideally then run externally around the office accommodation with junctions off as necessary to feed heaters etc.

As the supply is piped externally any leak in the system will normally be dissipated by normal air movement and not allowed to build up as it could if it was piped internally.

Whether supplied in bulk or by the more normal cylinder there are several points to be remembered:

1. Regulators, which are designed to control the pressure of the supply, should be fixed outside the buildings and supplies taken in at a low pressure to serve the appliances.
2. Internal piping should be rigid copper or iron as far as possible and flexible hose connections to appliances kept to a minimum.
3. Permanent and adequate ventilation must be provided. The products of combustion are carbon dioxide, nitrogen and water vapour, none of which is toxic. Continued burning in an unventilated room, however, will result in an oxygen deficiency arising and any occupants will become drowsy, may fall asleep and, if not disturbed, could become asphyxiated.
4. If the supply is provided by cylinder then the cylinder should be sited outside the building and turned off at the

Fig. 13.2 Bulk installation tank.

valve when not required for use. This ensures that any leak in the system is isolated and prevents any build-up of gas.

5. All flexible hoses must be properly clipped to ensure a leak-proof connection. This *does* apply to butane which tends to be known as low pressure and so not worthy of such attention.

Bitumen boilers

Boilers produce a great deal of radiant heat and it is very important that the supplying cylinder is at least 3 m away from the burner equipment, otherwise a pressure build-up can occur in the cylinder.

This longer-than-usual length of flexible hose is in itself susceptible to damage from site traffic and should be carefully positioned. The use of an armoured hose is also advisable. A further hazard with boilers is the possibility of the bitumen boiling over and careful supervision is necessary to check any tendency on the part of the operatives to overfill.

Hand-operated torches

Several trades find the small LPG bottle equipped with a blowtorch a very useful piece of equipment, but it must also be used with care.

The very portability of the bottles makes using them off platforms and tower scaffolds an easy proposition but if used in this way they must be properly secured to prevent them falling or being pulled onto their side. Hoses on this type of equipment should be inspected frequently as the constant flexing can cause wear particularly at the clipped ends.

One of the biggest dangers is when the operator uses the valve protecting cover on the bottle as a temporary rest for the lighted torch with the result that the flame plays onto the side of the bottle. This can in time create a hot spot and a build up of pressure sufficient to cause the bottle to rupture.

Fire control

The most effective form of fire control in respect of highly flammable liquids and LPG is fire prevention, but one must also be prepared for the emergency situation. Whenever highly flammable liquids are in use the correct extinguishing equipment must be readily available. Portable fire extinguishers, either Dry Powder, Carbon Dioxide or the 'multi-purpose' type, which is only an improved type of dry powder, should be provided *at the place of work* and not locked away in the general store.

Whenever LPG cylinders are involved in fire, either by exposure to a fire or by themselves feeding the fire, there is a real danger of a very severe explosion and this should always be borne in mind by the supervisor unfortunate enough to be faced with this situation. Fire control is dealt with more fully in Chapter 27 but a few ground rules may be useful at this juncture.

Fire from a leakage of gas

The most important, and in fact vital, rule in this situation is never to extinguish the flame except by turning off the cylinder valve, as a fire is in many ways to be preferred to an explosion. If it is possible to cool the cylinder, and other cylinders in the vicinity, by applying a water spray from a safe position then this should be done, but otherwise the area should be evacuated and the fire brigade called.

Cylinders exposed to a fire

When cylinders are exposed to a severe fire they are likely to explode in less than five minutes. Do not attempt to fight the fire but evacuate the area, call the emergency services and be sure to tell them of the presence of LPG cylinders in the fire area.

An indication of the potential of exploding cylinders is given in a report of a fire at a storage depot a few years ago. True this was a large store, much larger than the average site will be concerned with, but the effects could be similiar. As a result of exploding cylinders severe damage was caused to a factory and six houses and lesser damage to a further 127 houses and 12 industrial buildings. Most of the damage occurred within a radius of 180–230 m from the main fire area but one 7 kg cylinder fragment hit a chimney stack 533 m away, another 6 kg fragment damaged a roof 483 m away and an empty 50 kg cylinder landed on another roof 384 m away. This is the potential of that nondescript cylinder tucked away in the corner of your office. A very useful energy source, but one that needs your special attention.

14

Welding – gas and electric arc

The welding and cutting of metal is a process widely used on construction sites and it is important that the site manager is aware of the more common hazards, particularly those that are liable to affect the workforce generally as well as the welders themselves.

Welding of metal has been defined as being the union of pieces of metal at joint faces rendered plastic or liquid by the application of heat and the two common direct sources of heat used in the type of processes to be found on construction sites are:

1. Flame – produced by combustion of a fuel gas, generally acetylene or propane, with oxygen; and
2. Electric arc – struck between an electrode and the workpiece.

Both methods can produce hazards on sites, hazards individual to the method and also some that are common to both.

Flame welding and cutting

Identification of cylinders

Gas cylinders are painted different colours to aid identification. Oxygen cylinders are painted black. Fuel gas cylinders are painted maroon for acetylene and red for propane. As an additional safeguard, to prevent any interchange of fittings, the outlet valve threads of non-combustible gases are right-hand thread and those of combustible fuel gases are left-hand.

Storage of cylinders

Cylinders when not in use, whether full or empty, must be safely stored and it is preferable for stores to be provided in the open air, separating oxygen from the fuel gases.

Fig. 14.1 Cylinder store.

Cylinders stored in this way should be in a secure compound and protected from the extremes of weather, both the direct rays of the sun and from snow and ice. Typical arrangements for cylinder stores are shown in Fig. 14.1

Cylinders may be stored inside a building if it is not practicable to store them in the open air but in this case all fuel gases must be inside a fire-resistant store. They must not be stored with any other material or equipment and no naked lights, heating or smoking must be allowed in the store. Any lighting provided for the store must be flameproof, both lamps and switches, and ventilation is necessary both at high and low levels.

It will be seen from the previous paragraph that open-air storage is to be preferred not only on safety grounds but also for reasons of cost.

Oxygen cylinders may be stored lying down but not stacked more than four cylinders high; be sure that any cylinders so stacked are chocked to prevent rolling, but fuel gases must always be stored, and used, in an upright position as cylinders lying on their side with leaking valves can release liquid into the atmosphere. One volume of liquid produces approximately 200 volumes of gas almost immediately, so any leaks of liquid from cylinders must be avoided.

Transporting and handling cylinders

The transporting of cylinders onto and around construction sites needs careful organization and supervision if hazards are to be avoided. The valves on propane cylinders are provided with a guard, but those on oxygen and acetylene are more vulnerable and can easily be damaged by careless handling. Transport drivers should make use of mats or skids when unloading cylinders. Never permit cylinders to be tipped from dumper skips or dropped from lorries. The handling of individual cylinders is simplified by using canvas stretchers and this method decreases the possibility of the cylinder valve being used as a handhold.

Purpose made trolleys should be provided for pairs of bottles in use and the bottles should be secured to the trolley. The trolley serves two main safety features – it makes quick easy removal of the bottles simpler in an emergency situation and it ensures that the bottles are kept in an upright position, the importance of which has already been explained.

Lifting cylinders by crane

Extreme care is necessary when using cranes to lift cylinders. Never permit the use of chain or wire rope slings to lift single cylinders as the sling will inevitably slip along the cylinder. Rope, canvas and rubber slings may be used with great care but ideally properly constructed bottle holders or cages should be used. Cylinder trolleys may not be suitable and should be subjected to careful inspection before being stressed in this way.

Dangers of oil and grease

Oil or grease will ignite violently in the presence of oxygen and consequently oxygen cylinders must be kept away from all sources of oil or grease and not handled with oily or greasy hands.

Maintenance of equipment

As in all equipment regular maintenance of welding gear is most important. Cylinders and valves should be kept clean and hoses regularly inspected for damage.

If it is necessary for hoses to be joined then only standard connecting fittings must be used. In particular the manager should be aware of the danger of 'do-it-yourself' connections making use of copper tube, as copper, or a copper-rich alloy, coming into contact with acetylene can form an explosive compound.

Equipment in use

There are several important factors to be considered whenever gas welding equipment is being used.

Fire risks

Managers should be continually vigilant over fire risks, which can arise from several sources. Sparks from welding can travel a considerable distance and welding work on or near wooden floors or near wooden constructional work should be carefully controlled. Fire extinguishers should be readily available and the practice of fitting an extinguisher to the cylinder trolley is to be recommended. The multi-purpose extinguisher, which is an improved type of dry powder, is the approved type.

Carelessly directed torches are also a common source of fire, as are torches left alight and unattended, particularly when left with the flame directed onto the supplying cylinder.

A good safety practice with fire prevention in mind is to arrange for an inspection of the working site after any welding or cutting has finished to check for smouldering fires.

Hot work

If the welding area is accessible to other personnel, or to visitors, any completed work which is left hot should be plainly marked. This is good general practice and is followed by most experienced welders, but should be insisted upon. The fairly simple method of chalking 'HOT' on the workpiece is quite suitable.

Work on painted, plated or other surface treated materials and the use of fluxes

Ventilation of the working place is always important when using gas welding equipment but is vital when the work involves painted or plated surfaces or the use of fluxes. Galvanized and cadmium plated surfaces are particularly dangerous and local exhaust ventilation should be provided when welding or cutting even in well-ventilated workplaces. In naturally poorly ventilated areas breathing apparatus may be preferable. Care should be taken, when positioning any exhaust ventilation equipment, that the outlet disperses the extracted fumes in a safe open air area.

Welding in enclosed places

Particular care is necessary if a welder is required to enter a confined space to perform a welding or cutting operation. Never permit the gas cylinders to be taken into the confined space but station a second man, who understands the equipment, outside to control the gas supply and provide any other assistance that is necessary. The blowtorch must always be lit outside the confined space and passed in to the welder. A special check should be made of the hoses and connections as any leak either of oxygen or of the fuel gas could prove to be very dangerous.

Ventilation is also, of course, of prime importance and must be maintained at all times, but oxygen must never be used for ventilation purposes. There have been cases when welders, not appreciating the dangers of working in an oxygen-enriched atmosphere, have used an oxygen line to clear fumes in a confined space. Their clothing, probably a little greasy, has been ignited by a spark with fatal results.

Operator's personal protection

Protection from sparks and hot slag, particularly for hands, legs and feet, and suitable eye protection is essential for operators using gas welding equipment. Close-fitting overalls topped by a leather apron, protective gauntlets and welding goggles of the right filter quality are minimum requirements.

Emergency procedure

If an acetylene cylinder is exposed to an external source of heat or becomes hot due to severe backfires, the operator, if properly trained, will be aware of the action he should take. The manager should also be conversant with the emergency procedure which is as follows:

Shut the cylinder valve, detach the regulator and hose and take the cylinder into the open air, well clear of any sources of ignition. If possible immerse the cylinder in water or alternatively apply water to it in large quantities. Open the valve and continue to cool with water until the cylinder is empty. All personnel not directly involved should be cleared from the area, warning notices posted and the Fire Brigade advised.

Electric arc

During electric arc welding the electrode and part of the holder are electrically live and hot so that if they are touched electric shock or burns may be sustained. Additionally the arc gives off a high intensity light, infra-red and ultraviolet radiation, while fumes arise from the process.

Power sources

Most arc welding on construction sites will use locally generated electrical supplies and diesel or petrol engined generators are regularly used for this purpose. These generators can introduce their own potential hazards and must be carefully sited to ensure that there is no danger from a build-up of exhaust gases.

Voltages for hand-held welding operations are relatively low but are sufficient to cause an operator to fall or drop something if a shock is received.

Wet and damp underfoot conditions increase the danger of electric shock. If possible the welding leads should be kept off the ground under wet conditions to help reduce this possibility.

Personal protection

Protective clothing and headgear is essential for the operator and no part of the body should be exposed to arc rays. The most important part of the protection is the head shield which can either be hand-held or an integral part of a helmet. The same protection rules apply to any assistants working with the welder.

'Arc eye', sometimes known as 'eyeflash', can be caused by looking at an unscreened arc for a few seconds only and even from a distance of some metres.

The condition does not appear immediately but usually between 4–8 hours after the exposure and the symptoms are a very unpleasant feeling of sand in the eyes, pain of varying intensity, an intolerance to light and occasionally temporary loss of vision. The condition will generally clear up between 24–48 hours but is very painful and to be avoided.

It will be appreciated that this particular hazard can affect passers-by or other employees working in the vicinity and the importance of screening around arc welding operations cannot be over emphasized. The screens should be of non-combustible materials and non-reflecting. Matt black is the ideal finish for screens of this sort. It is good practice to post notices warning passers-by that arc welding is taking place.

Fumes and metal fume fever

The importance of good ventilation has already been discussed in relation to gas welding processes and is, if anything, more necessary when arc welding processes are in operation. Particular care must be taken when working on galvanized or cadmium plated metals and there must be a high standard of ventilation and fume extraction.

Metal fume fever is, as is arc-eye, a short but acute illness and is brought about when fumes, produced from a metal heated above its melting point, are inhaled. It is not, of course, limited to welders but is fairly common amongst them. The symptoms are a foul taste in the mouth, irritation in the nose and throat and coughing followed by muscular pains, weakness, fatigue and a general feeling of being unwell. The symptoms are very similar to, and sometimes confused with, an influenza-type cold. Approximately 10–12 hours after the exposure fever or chills or both are noted and these may be quite severe, with chilling lasting as long as three hours in some cases. There is no chronic form of this particular disease but in rare instances complications such as bronchitis may follow.

The treatment for this illness is limited as most cases are well on the way to recovery within 24 hours, but keeping warm in bed and drinking plenty of fluids is recommended. The best treatment, as in most illnesses, is prevention.

Tidiness and good housekeeping

Good housekeeping and the removal of all litter, rubbish and combustible material from the work area is one of the cardinal points of site management but never more important than when in connection with a welding process.

The everpresent fire source can easily lead to the outbreak of a fire of materials in the locality if proper care is not taken. If welding is required in an area where non-movable combustible material is located then arrangements must be made to cover this material with non-combustible sheeting or boards.

Welders, because of the protective clothing and equipment they wear, have a restricted field of vision and a non-cluttered work area is most important for them. They obviously have a responsibility in this matter themselves but are only partly able to control their working environment. The site manager must ensure that a proper lead is given to ensure safety in this particularly hazardous operation.

15

Working with asbestos

Asbestos is the name used to describe a group of silicates which occur naturally in fibrous form. It was first used in Finland about 2500 BC to strengthen clay pots, historically appears in indestructible woven shrouds for preserving the ashes of the famous, has served as lamp wicks through the ages up to the present day and currently has many uses in various branches of the construction industry.

Types of asbestos

There are four main types of asbestos, namely:

1. Chrysotile: a fine, silky white/grey fibre known as 'white asbestos'.
2. Crocidolite: a straight, flexible blue fibre, known as 'blue asbestos'.
3. Amosite: a straight brittle grey/brown fibre, known as 'brown asbestos'.
4. Anthophyllite: a brittle white/brown fibre.

By far the commonest type of asbestos is chrysotile, which accounts for approximately 90 per cent of the total produced annually and which is widely used in the production of asbestos cement sheeting, pipes, flooring tiles and asbestos woven products.

The import of crocidolite (blue asbestos) in its raw state has been banned since 1970 but this type of asbestos, accepted as being the most hazardous, is present in many existing buildings and installations as thermal insulation or sprayed coatings.

Amosite is used in the production of asbestos insulation board which is widely used where thermal insulation and fire protection are important.

Anthophyllite is not widely used in the construction industry although it may be present mixed with other types, and for our purpose can be linked as a hazard with chrysotile.

It is usually difficult to determine from its colour which form of asbestos is present in a manufactured product.

The examples given above are only a guide to the possible type of asbestos involved and it should be pointed out that crocidolite cannot always be identified by its 'blue' colour because of the browning effect of heat.

The only certain method is to have the material analysed by X-ray diffraction or microscopy.

Health hazards

Materials containing asbestos present a hazard only when asbestos fibres are released and dispersed so that they can be inhaled. When asbestos dust is inhaled fine particles, which are much too small to be seen by the naked eye, are deposited in the lungs and may cause one of the following conditions:

Asbestosis

This is a scarring and thickening of the lungs and mostly occurs after many years of exposure to high concentrations of asbestos dust. The characteristic symptom is a progressive breathlessness and a hard unproductive cough. Asbestosis is not of itself a cause of much disability but it has been shown that there is an increased incidence of cancer of the bronchus or lung in people who are already suffering from asbestosis. It has also been proved that the risk of getting lung cancer from inhaling asbestos is many times more serious if one smokes cigarettes. Smoking should be avoided by anyone who is exposed to asbestos dust regularly.

Mesothelioma

This is a tumour of the lining of the chest and in some cases the lining of the abdominal cavity and is usually associated with heavy and prolonged exposure to crocidolite (blue asbestos).

All asbestos related diseases may, and usually do, occur some years after the person has been exposed. There is often no obvious risk at the time of exposure and consequently the need for supervision, instruction and education is greater than for more visible risks to health.

Site procedures and obligations

The Asbestos Regulations 1969 require every contractor and every employer who is undertaking work involving an asbestos process to comply with the requirements of the regulations, both for his own employees and also for any other persons employed there and who may be liable to be exposed to 'asbestos dust'.

'Asbestos dust' is defined in the regulations as being dust consisting of or containing asbestos to such an extent as is liable to cause danger to the health of employed persons.

As already described, the identification of asbestos types can only be done by specialist analysis and similarly the measurement of concentration of asbestos in the atmosphere must be obtained by using a sampling instrument to decide whether special working precautions are required. In the working environment of a stable factory-type situation regular monitoring of dust concentrations is a relatively easy matter but on the ever-changing construction site scene the testing process is less easy, particularly when one appreciates that testing is normally carried out over a four-hour sampling period and that the filters, once obtained from the samplers, have to be laboratory analysed before a pronouncement can be made.

It is obviously desirable, therefore, that some basic guidelines be given but it must be emphasized that measurement is the only way to quantify dust concentrations and that the examples following of common operations and the probable risks resulting can only be of a general nature.

First let us consider the materials concerned.

Asbestos-cement

Asbestos-cement sheeting, pipes etc. contain approximately 10–15 per cent by weight of asbestos fibres. This is considered to be a fairly low percentage content and although the asbestos content alone does not give a complete indication of the risk involved it can be a useful guide. Accordingly little danger exists when carrying out simple hand operations on asbestos-cement products. Even small-scale machine sawing or drilling can be undertaken in the open air without undue risk. However, work on asbestos-cement products in a more confined space using machine saws should not be undertaken without the use of dust extraction equipment.

Asbestos insulation board

The proportion of asbestos in asbestos insulation board is higher and can be as high as 40 per cent of the total weight. It naturally follows that work associated with machine sawing, shaping and drilling of this material creates a greater hazard and dust extraction equipment must be used, particularly when drilling overhead. The provision of this extraction equipment can be costly and consideration should be given to arranging, wherever possible, for cutting and drilling of boards to be carried out off-site under factory conditions. Remember that, as has already been mentioned, materials containing asbestos are not a hazard in themselves but only when asbestos dust is released into the atmosphere, so the fixing of clean pre-cut, predrilled panels presents no problem (see Fig. 15.1).

Sprayed thermal insulation and lagging in heating systems

For some considerable period asbestos has been applied by spray process as a fire protection for structural steelwork, to walls and undersides of roofing to provide acoustic and heat insulation and

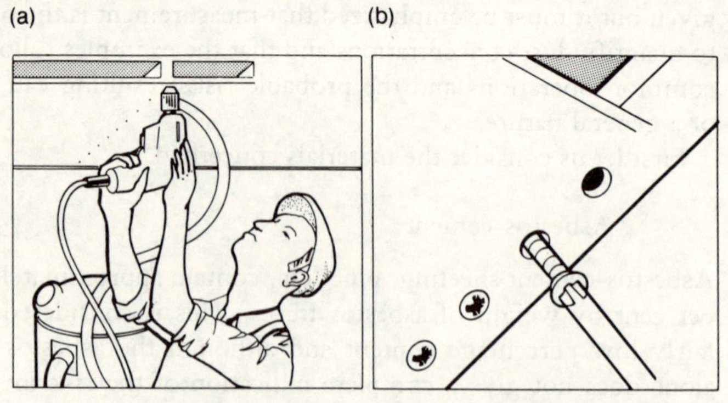

Fig. 15.1 Overhead drilling and fixing of pre-drilled boards. (a) Where overhead power drilling is being done, portable dust extraction equipment must be used. (b) Better still, boards, panels or tiles for fixing overhead should be pre-drilled and fixed with TEKS screws.

as lagging in heating systems. Prior to 1970 most of these applications will have contained blue asbestos (crocidolite) and the coating will normally contain a minimum of 55 per cent asbestos with Portland cement as a binder.

The removal and/or repair of insulation of this type presents a real hazard to the contractor and requires careful planning and precise adherence to the requirements of the Asbestos Regulations 1969.

Not least in importance for the site manager to remember is the requirement to give not less than 28 days written notice to the District Inspector of Factories of the intention to start any work involving the removal of lagging consisting of or containing crocidolite.

The Inspector will satisfy himself that the work arrangements comply with the other requirements of the regulations and, if indeed crocidolite is present, will require the operatives to wear positive-pressure respirators and full protective clothing.

Whether crocidolite is present or not the removal of thermal insulation needs great care, and the advantages of thoroughly wetting the lagging before commencing removal cannot be overemphasized, considerable reductions in dust being achieved.

If at all possible the area to be stripped should be enclosed with polythene sheeting and a warning notice displayed. This will help to prevent the spread of the dust, but the site supervisor is strongly advised to bring in a company experienced in this type of work whenever stripping of asbestos is required in an area that cannot be isolated. (see Fig 15.2).

The remaining points in the regulations that should be noted are:

1. Exhaust ventilation must be provided to any mechanical equipment used for working on asbestos to prevent asbestos dust entering any workplace and the exhaust ventilation equipment so provided must be examined at least once in every seven days and thoroughly examined and tested once every fourteen months. The machine sawing and drilling of insulation board on site would require this precaution.
2. If for any reason it is impracticable to meet the requirements regarding exhaust ventilation or to prevent dust effectively by damping, then respiratory protective equipment and protective clothing must be provided. Operatives using the respirators must be trained in the use of the

Fig. 15.2 Wetting asbestos and enclosures. (a) Wherever possible, *any* asbestos material should be thoroughly wetted before stripping. (b) Areas to be stripped should be enclosed with polythene sheeting and a warning notice displayed.

equipment and are obligated to use it.

3. Cleaning is also very important and the regulations require that all machinery, floors, working surfaces, etc., should be kept clean and free from asbestos waste and dust. If possible, stripped material should be placed directly into polythene bags and not allowed to create dust by dropping on the ground. Any cleaning that is required is to be carried out by a dustless method, that is, either the dust should be damped down or suitable industrial vacuum cleaners used.
   If crocidolite is included in any waste bags then the bags must be clearly marked 'Blue asbestos – do not inhale dust'. (see Fig. 15.3)
4. Suitable accommodation should be provided for changing into and from protective equipment. When protective clothing is sent for cleaning it should be sealed in dustproof containers and clearly marked 'Asbestos contaminated clothing'.
5. Young persons under the age of 18 must not be employed in any asbestos process requiring the use of respirators and protective clothing.

Disposal of waste

The deposit of asbestos waste and dust is controlled by the Control of Pollution (Special Waste) Regulations 1980 which places duties

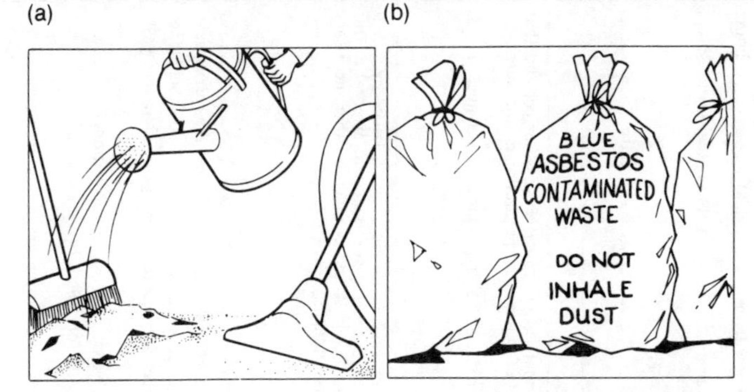

Fig. 15.3 Cleaning and bagging. (a) Always thoroughly wet before sweeping. (b) Stripped material and floor clearings must be put into dust-proof polythene bags securely fastened for properly authorized disposal. Blue asbestos waste must be clearly labelled.

on producers, carriers and disposers of special waste. A study of the complete code of regulations is recommended but briefly the duties of the 'producer', that is in our case the site supervisor, are as follows:

He must prepare six copies of a consignment note (see Fig. 15.4) in the approved form and pass one copy to the disposal authority for the area in which it is to be disposed of not less than three clear days before the removal of the waste. He will, at this stage, only complete Parts A and B of the consignment note. The person who collects the waste will complete Part C of the note and the site supervisor will then complete Part D, retain one copy of the note and pass the remaining copies to the collector. The collector, or carrier as he is described in the regulations, will be responsible for passing the copies of the note to the disposer, or tip manager, for final completion.

The system appears rather cumbersome but does ensure that the local disposal authority is given both notice of the intending disposal and also confirmation that the disposal has been carried out. This close control is essential in ensuring the safety of the general environment of our country.

Department of the Environment/Welsh Office/ Scottish Development Department CONSIGNMENT NOTE FOR THE CARRIAGE & DISPOSAL OF HAZARDOUS WASTES	Serial No.

Producer's Certificate **A**	(1) The material described in B is to be collected from .... and (2) taken to .... Signed .... Name .... On behalf of .... Position .... Address and telephone .... Date .... Estimated date of collection ....
Description of the Waste **B**	(1) General description and physical nature of waste (2) Relevant chemical and biological components and maximum concentrations (3) Quantity of waste and size, type and number of containers (4) Process(es) from which waste originated
Carrier's Collection Certificate **C**	I certify that I collected the consignment of waste and that the information given in A(1) & (2) and B(1) & (3) is correct, subject to any amendment listed in this space: I collected this consignment on .... at ....:.... hours Signed .... Name .... Vehicle Registration No. .... On behalf of .... Address and telephone .... Date ....
Producer's Collection Certificate **D**	I certify that the information given in B & C is correct and that the carrier was advised of appropriate precautionary measures. Signed .... Name .... Telephone .... Date ....
Disposer's Certificate **E**	I certify that Waste Disposal Licence No. .... issued by .... Council, authorises the treatment/disposal at this facility of the waste described in B (and as amended where necessary at C). Name and address of facility .... This waste was delivered in vehicle (Reg. No.) .... at ....:.... hours on (date) .... and the carrier gave his name as .... on behalf of .... Proper instructions were given that the waste should be taken to .... Signed .... Name .... Position .... Date .... on behalf of ....
For use by Producer/ Carrier/ Disposer	

Fig. 15.4 Specimen form from Schedule 2 of The Control of Pollution (Special Waste) Regulations 1980

16

Site transport and mobile plant

The safe use of transport and mobile plant on construction sites is largely a matter of taking commonsense precautions but there are some legal obligations that need to be met and which reinforce the commonsense approach.

Legislation

The Construction (General Provisions) Regulations 1961 has a small section which deals with transport generally and the regulations most concerning us are:

Regulation 32 which requires that mechanically propelled vehicles must be driven only by trained men over 18 years of age. Provision is made for men under that age to drive whilst under training but they must be directly supervised by a qualified driver.

Regulation 34 which requires that all vehicles and trailers used on a construction site for carrying men and materials must:

1. Be in good working order and be maintained in good repair.
2. Not be used in an improper manner.
3. Not be over loaded or unevenly loaded.

Regulation 35 which prohibits the carriage of passengers in vehicles where no provision for carrying passengers has been made.

Regulation 36 which prohibits persons remaining on a vehicle which is being mechanically loaded with loose materials.

Regulation 37 which requires that measures must be taken to prevent vehicles running over the edge of excavations etc. when tipping materials.

These regulations, like all of the other regulations in the various construction codes, were born of necessity and are a good starting point for the supervisor seeking to maintain a good safety standard in respect of transport on site.

The types of vehicles that we are considering include trucks, tipper lorries, tractors and trailers, dumpers and rough-terrain forklifts. There are precautions necessary that are applicable to them all and also specific hazards for individual types and it is sensible that we look at the more important aspects in greater detail.

General precautions

Driver selection and training

Technically a driving licence is not required to drive a vehicle on a contract site but if the public highway is used or crossed then the appropriate licence is necessary. It is obviously desirable, however, that any prospective site driver does, in fact, hold a current driving licence as this at least indicates a minimum standard of driving competence, but the importance of training cannot be stressed too greatly and the supervisor has the responsibility, under regulation 32 discussed previously, to see that every driver on his site is trained in operating his vehicle. Having provided the training the next sensible step is to issue all trained drivers with written authority and to prohibit any untrained operative from driving a site vehicle. This will need strong supervision but is essential. A precautionary word about the age of drivers may be helpful. We have already discussed that the minimum age for site driving is 18 but when vehicles are used on the public highway other rules come into effect. The most important of these is the requirement for a minimum age of 21 years when the permissable weight of the vehicle exceeds 7.5 tonnes in the case of engineering plant and 3.5 tonnes in the case of dumpers.

Site conditions

We have already briefly discussed in Chapter 3 the need for planning the site roads and access points, and many site transport accidents could be avoided if more thought was given to providing a safe route for traffic. Congestion of work and stores areas can create problems for drivers required to reverse their vehicles; routes under or adjacent to overhead lines can be hazardous and, if unavoidable, must be properly signed and the cables protected; mud, lying water and icy conditions all add to the hazards and must be kept under control; and routes requiring vehicles to travel across steep slopes should be avoided, as traffic moving in this way is very susceptible to overturning.

Protection for the edges of excavations and banks is also very necessary as vehicles are regularly involved in accidents on site through either driving into an excavation or being allowed too close to the edge and causing a collapse of the excavation wall. This is, of course, particularly necessary when lorries or dumpers are tipping into excavations and in such cases adequate stops are required by Regulation 37. Baulk timbers, securely anchored, are quite usual and satisfactory for this job providing they are of sufficient size. At least 300 mm square timbers should be used and care must be taken to ensure that the timbers are moved and relocated as necessary when the work proceeds. A very good alternative idea to baulk timbers has been around for some time and consists of a stop fabricated from steel tube or angle (see Fig. 16.1).

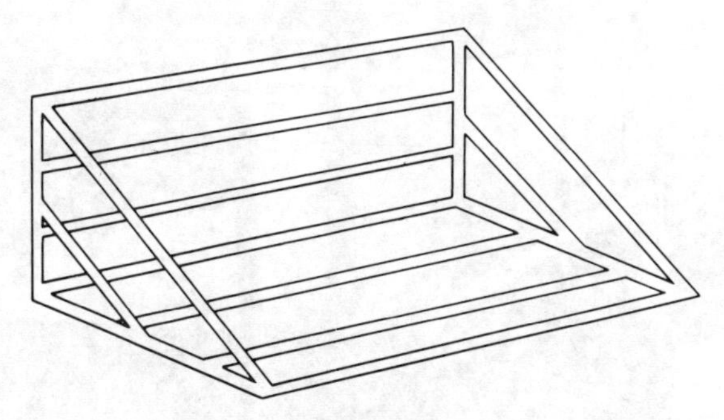

Fig. 16.1 Vehicle stop.

The principle is that two units are used, one for each wheel of the tipping vehicle nearest the excavation edge, and the weight of the vehicle anchors the stop. The use of two units is to be preferred to one larger one as handling is easier and consequently the necessary moves of the stops are more likely to be made.

Tipping lorries – special hazards

Many accidents involve people being struck by reversing vehicles when the driver's rear view is obscured. Drivers must be given firm instructions that they must use the services of a banksman who can give clear signals. In the unlikely event of there not being a banksman available the driver must not reverse until he has satisfied himself, by alighting from his cab if necessary, that no one is at risk.

Other regular accidents involving tipper lorries are those where the tail-gate, whilst being opened or closed, strikes or traps an operative, and those where the drivers slip from the sides of the body when carrying out sheeting operations. Some enlightened companies are designing these hazards out of the lorries by providing pneumatic tail-gates and by fixing walkways along the lorry sides, but it seems to be slow to catch on and the only alternatives are extreme care on the part of drivers and operatives and regular checks to ensure that latches on tail-gates are performing correctly (see Fig. 16.2).

Fig. 16.2 Tail-gates and walkways.

Tractors and trailers

In the author's experience the main hazard to be found with tractors and trailers on construction sites is that they seem to be considered as items of communal transport, to be driven by every Tom, Dick and Harry, given minimum maintenance and still expected to provide safe transport. If your driver authority system is efficient then you may not have this problem but in any event there is one important feature of agricultural type tractors, and there are many used on construction sites, of which you should be reminded. This type of vehicle is provided with two brake pedals which enable the

Fig. 16.3 Tractor pedals.

driver to stop the rotation of either the left or right rear wheel, depending on which pedal he uses. It is designed to facilitate rapid turns, such as the farmer needs when ploughing (see Fig. 16.3).

There is a bridge piece which slots across the pedals and permits the brakes to be applied to both wheels and this is where the danger lies. There have been many cases of an inexperienced driver, not appreciating this single wheel braking facility, being caught off guard, veering off course and meeting with a serious accident. If you have a tractor of this sort on site and the work it is required to do makes this very sharp turning unnecessary, be extra safe and fix a temporary bolt to make sure the pedals are always bridged.

Dumpers

Dumpers are the most widely-used vehicles for transporting materials on site and, although they are fairly straightforward to operate and maintain, accidents happen frequently, caused by careless operation, lack of maintenance or the carriage of unauthorized persons.

Mud on the control pedals is a regular contributing factor, allowing the foot to slip off the pedal with varying results, and there have also been cases of pedals being inoperative because of a build-up of mud and concrete waste.

Dumpers left standing with the engine running are also a hazard. Many cases of machines being knocked into gear by careless loading are reported and many serious injuries result.

The carriage of unauthorized passengers on dumpers, particularly small dumpers, is very common on sites unless the supervision is very alert. Each machine should display a notice prohibiting such activities and disciplinary action should be taken against any defaulters.

Rough-terrain fork lifts

This type of fork lift has been specially produced for use on construction sites and has many differences from the older-established type of fork lift 'truck' used in the more stable factory and depot situation. It is therefore very important that only rough-terrain fork lifts are used on construction sites and additionally appreciated that 'rough-terrain' does not mean 'mountainous'.

Safe operation

With the right machine for the job in hand and reasonable site conditions the safe operation of fork lifts on site still depends very much on correct operation.

The following are examples of unsafe actions that have led to, and will inevitably again lead to, accidents involving fork lifts:

1. Persons standing under an elevated load.
2. Lifting a load greater than the machine rated capacity.
3. Lifting a load on uneven ground with the mast out of vertical.
4. Not positioning the forks fully under the load and at the correct spacing.
5. Travelling with a load and the mast vertical – it should be inclined backwards.
6. Travelling with a load in a raised position.
7. Raising or lowering the load whilst travelling.
8. Not checking the stability of the load.
9. Descending ramps or slopes with the load in front.
10. Discharging a load from inertia force, i.e. applying the brakes harshly.
11. Travelling without a load with the forks raised more than necessary to clear the ground.
12. Driving over rough terrain at speed.
13. Driving across a slope and exceeding a safe working angle.

14. Cornering at speed.
15. Braking violently or making sudden stops.
16. Not facing or observing the direction of travel.
17. Failing to observe overhead clearance.
18. Failing to observe 'rear end swing' especially when manoeuvring in confined spaces.
19. Subjecting scaffolds or loading towers to unnecessary shock loads.
20. Loading scaffolds or loading towers beyond their rated capacity.
21. Using the machine as a working platform without proper safeguards.
22. Carrying passengers.
23. Failing to park the machine correctly, e.g. failing to:
    (a) position machine on level ground;
    (b) position forks flat on ground;
    (c) switch off engine;
    (d) apply brakes;
    (e) set controls in neutral;
    (f) remove ignition key.

Driver training should help to remove most of the hazards but only if supported by good supervision.

Vehicle maintenance

The importance of maintenance has already been mentioned several times in this chapter and there is no doubt that the lack of, or a poor standard of, maintenance leads to many accidents.

The supervisor should satisfy himself that a systematic maintenance routine is in being. Basic maintenance can usually be carried out on a daily and weekly basis by the drivers, who must be instructed to report any defects immediately, but a regular check by experienced fitting staff should also be included. Ensure that your drivers are conversant with the maintenance routines and hazards and never permit them to attempt repair work in which they are inexperienced. Many experienced fitters sustain accidents when repairing plant, even in ideal workshop conditions, and repairs carried out on site are always potentially more hazardous.

17

Static mechanical plant and equipment

It has been said before and, no doubt, will be said again, that safety regulations are only imposed when the accident record of a particular branch of industry shows that some form of regulating legislation is necessary.

There is no reason for static mechanical plant to differ in this and accordingly we find that a special code of regulations exists for woodworking machines. On consideration, this is not too surprising as many of the machines in that category are obviously capable of inflicting frightening injuries and woodworking machines have been on the scene longer than most other items of plant. However, it does tend to separate 'woodworking machines' from other static plant such as concrete mixers, pumps, silos and so on and accordingly we shall follow that example.

Woodworking machines

The Wood-working Machines Regulations 1974 give comprehensive instructions and guidance on the general operation of a wide range of woodworking machines.

Definitions

The regulations define woodworking machines as: all types of circular saws; band saws; chain saws; grooving, mortising and planing machines; spindle moulding machines; multicutter moulding machines; tenoning machines; trenching machines; automatic and semi-automatic lathes; and boring machines. In fact it is fair to say that any machinery that is used to cut, drill or shape wood, whether fixed machinery or hand-held, is included.

Obviously very few construction sites will rate more than a circular saw, and we shall concentrate on those, but the extent of the definition should be appreciated and study made of the complete code of regulations if the setting up on site of a small joinery shop is envisaged.

Circular saws

Guarding

Saw blades must be guarded below the bench and provided with a strong adjustable top guard which must be constructed with side flanges and, when in use, be adjusted as close as possible to the material being cut.

A riving knife must be fitted in line with the saw blade and must be curved to match the largest diameter saw blade that can be fitted. It must always be adjusted as close as practicable to the blade and never be more than 12 mm away at bench level. The height of the riving knife above the bench must be at least 225 mm when used with blades of 600 mm diameter or more and not more than 25 mm below the top of the blade when smaller diameter blades are being used. As well as guarding the back edge of the blade the riving knife also prevents timber from closing onto the blade and the adjustments referred to are particularly important (see Fig. 17.1).

Sizes of blades

Limitations on the sizes of circular saw blades are quoted in the

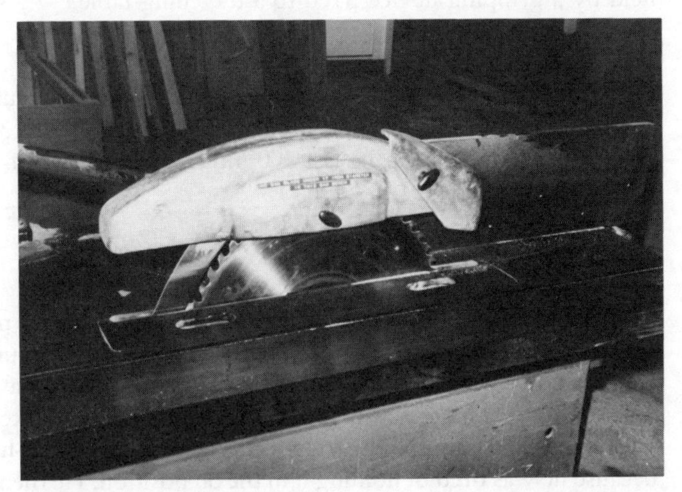

Fig. 17.1 Guarding circular saw.

regulations and the diameter of the blade in use must never be less than six-tenths of the diameter of the largest blade that can be fitted.

The regulations also require that a notice showing the minimum diameter of saw blade to be used is fixed to every saw bench. This is important 'information' and one which supervisors would do well to check.

Legal limitations on the use of circular saws

It is possible with many types of circular sawing machines to raise and lower the blade so as to vary the height of blade above the bench and this facility has tempted operatives to use the machine to cut rebates, tenons and grooves. This can be a very dangerous operation, as it obviously necessitates the removal of the standard top guard, and the regulations do prohibit the use of a circular saw for this purpose unless the upper part of the saw is effectively guarded. Special guards, completely boxing in the blade apart from a feeding and a removal opening, can be designed to comply with the regulation, but this is the only permitted method and even then the requirement for a properly adjusted riving knife, as previously discussed, still remains.

This particular regulation goes on to require that the saw teeth must always project *above* the material being cut, except when cutting rebates etc. as described above, and also prohibits the cross-cutting of logs or branches unless the material being cut is firmly held by a gripping device fixed to a travelling table.

Push sticks

Suitable push sticks must be provided at all fixed circular saw machines and must always be used for feeding timber of 300 mm or less in length and on the last 300 mm of longer lengths of timbers, as seen in Fig. 17.2.

Hand-held circular saws

This portable version of the circular saw warrants special mention because of the regularity with which it is involved in injury accidents. Faulty switches or, worse still, switches that have been tampered with, are a case in point. One recorded account tells how a carpenter had jammed the trigger mechanism on a hand-held saw because he was tired of holding it in the on position. He then turned the electric supply off at the wall socket switch and laid the saw

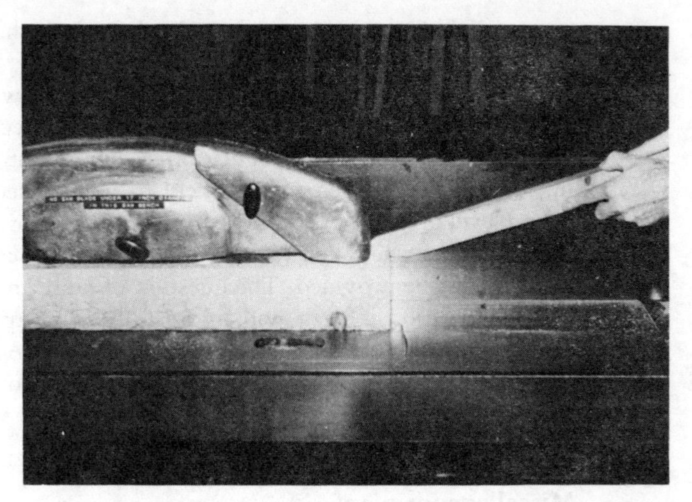

Fig. 17.2 Push stick in use.

down on his work bench to go off for his tea break. Returning after his break he unthinkingly turned on the power at the wall socket and the portable saw immediately came to life.

The saw blade bit into the bench and the whole machine was propelled across the workroom striking another carpenter and causing severe lacerations. Other serious injuries have resulted from guards being removed or jammed. Blade guards on this type of machine are sprung, to ensure that only a minimum of blade is exposed, and the effectiveness of these guards must be regularly checked.

General requirements

Several general safety points apply to all woodworking machines:

1. An easily accessible stop control must be provided and clearly marked.
2. Sufficient space must be maintained around machines so that the work can be done safely and the floors or ground area must be level, not slippery, and be kept clear of wood chips, offcuts etc.
3. Adequate and suitable lighting must be provided and suitably shaded in the case of artificial lights.
4. Except when machines are sited in the open air, a reasonable temperature of 13°C must be maintained or, if this is

not practicable, heaters must be provided where the operatives can warm themselves.

5. Finally the question of noise arises and the woodworking machines code of regulations is the first to deal specifically with this problem. The regulation requires that where operatives are exposed to, or likely to be exposed to, a sound level of 90 dB(A) or more for eight hours a day then measures must be taken to reduce the noise and ear protectors must be provided. The question of noise and hearing conservation is dealt with more fully in Chapter 24. Suffice it to say at this point that a level of 90 dB(A) is not very high in terms of woodworking machine noise and great care must be taken when such machinery is sited.

Miscellaneous plant

The most regular causes of injuries connected with plant are:

1. Lack of machine guards.
2. Lack of operator training.
3. Poor machine maintenance.

Machine guards

Every moving part has to be securely guarded unless it is absolutely safe by position and the common starting-dog shaft is the most regular culprit. Gradually items of plant are being provided with electric starters but many of the smaller mixers, pumps, etc., are still started by hand. A well-designed item of plant will have a dog shaft shrouded by the main machine cover, but other less thoughtful manufacturers provide a starting shaft that protrudes from the machine cover. The 'afterthought' guard provided for this shaft is regularly damaged, broken off or removed and the supervisor should be watchful for an exposed shaft which can easily pick up loose clothing and cause serious injuries.

Operator training

A lack of training and instruction, particularly in hand-starting techniques, is another major accident cause. There was a time when most motor cars were provided with a starting handle and in consequence a good many of us had been taught, or had learned the hard way, the techniques of 'winding up' the petrol engine. Starting handles on private cars are now very much the exception and this starting skill is no longer an important part of car ownership. Operatives *do* need to be instructed in the starting techniques for both petrol and diesel engines and the difference in methods should be made absolutely clear to them.

1. *Petrol engines.* Make sure the starting handle is fully engaged. Grip the handle firmly, being sure to keep the thumb and forefinger together. Pull the handle UP a quarter of a turn – never push it down.
2. *Diesel engines.* Here quite a different technique is needed. Set the fuel pump control in the overload position. Lift the decompression lever, which is usually situated on top of the rocker box. Ensure that the starting handle is fully engaged and crank the engine, starting slowly and building up speed. Crank until the flywheel is turning reasonably fast and then drop the decompression lever. The engine should then start. Never *stop* a diesel engine by lifting the decompression lever as engine damage may follow. Always stop the engine by using the fuel pump control (see Figs. 17.3 and 17.4).

Machine maintenance

There is a tendency for items of plant that do not require an individual operator to be put at the bottom of the list when maintenance schedules are considered. This is quite the wrong approach

Fig. 17.3 Decompression lever use.

Fig. 17.4 Decompression lever use.

and regular inspection and maintenance of this type of plant and equipment is very important. Badly serviced plant quickly becomes difficult to start and, as well as leading to accidents, can cause expensive delays. Wire ropes on mixer skips need regular inspection, as do the rope anchors and pulley gear, and particular attention should be paid to the effectiveness of safety chains fitted to mixer supply skips.

Cement silos have brought advantages to sites by reducing physical handling of cement and subsequent reductions in back strain and dermatitis, but can also create problems of their own. The supervisor must ensure that proper precautions are taken if it is ever necessary for a person to enter a silo to carry out work. Two men must always be in attendance, a belt or line must be worn by the man entering the silo and adequate means of hauling him out in an emergency must be provided.

'Permit to work' systems are discussed more fully in Chapter 23 and the entry of a person into a silo is a perfect example of a sensible use of a permit to work, so ensuring that every care is taken.

18

Overhead and underground services

Every year there are many accidents to operatives on construction sites involving overhead electric lines and underground electric cables and gas supplies. The accidents are usually:

1. A result of machines or equipment coming into contact with live overhead electric lines or buried cables.
2. Caused by machines fracturing high pressure gas mains.

The cables and main supplies of the other statutory undertakings are not generally hazardous but they can be very expensive if damaged, as anyone who has dug through a transatlantic telephone cable will tell you! From the safety angle, however, we shall deal with electricity and gas and, although not of primary importance in the 'safety' context, any disruption of major supplies of these two services can also lead to considerable compensatory costs brought about by loss of production etc., a point which the cost-conscious supervisor will wish to bear in mind.

Our first concern, however, must be with the potential hazards that can arise when these services are damaged during construction operations. One important point needs to be made right away. Always assume that any overhead lines, and/or underground services, discovered are live. To do otherwise courts disaster.

Overhead electric lines

Overhead conductors are normally uninsulated and if contact or near contact is made with them by a crane jib, scaffold pole or similar metallic object, an electric current will discharge to earth through the crane or pole with the risk of fatal or severe shock and burns to any person in the immediate vicinity. This type of accident produces a very high incidence of fatalities (one in three results in death) but can be prevented if the necessary precautions are taken.

Planning the contract

When setting up his work programme the site supervisor should find out whether overhead lines pass over the site or any access point and, if this is the case, he should make contact with the Electricity Board for the locality and agree with the Board's Engineers a plan for a safe method of work. This approach should be made as early as possible as, if sufficient notice is given, it may be possible in some instances for the lines to be re-routed or to be raised to a safe height. Other alternatives can involve the lines being made dead for certain periods to allow work to proceed safely. A 'permit to work system' should be used in cases such as this. See Chapter 23 for details of 'permit to work' systems.

If neither of these alternatives can be arranged then barriers must be provided as required by Regulation 44 of the Construction (General Provisions) Regulations 1961.

Physical barriers

It is important that all concerned with the design and placing of barriers should fully understand that the purpose is not solely to prevent actual contact with an overhead line but also to ensure that all metallic objects remain at a safe distance. This 'safe' distance will obviously vary depending on the voltage being carried by the line and advice from the local Electricity Board must *always* be sought.

When working on site there are three main types of work operation that need to be considered:

1. Where there will be no work done under the lines nor any traffic or plant passing under the lines. These situations only require barriers to prevent inadvertent close approach.
2. Where plant needs to pass under the lines but no work will take place. The need here is for a clearly defined accessway.
3. Where work must be done beneath the lines. This situation obviously requires additional precautions.

No work or passage of plant

Ground level barriers should be positioned parallel to the lines and at a minimum distance of 6 m from the lines. The Electricity Board may require a greater distance, depending on the voltage concerned. Measurements should be taken at ground level, that is from a point underneath the overhead conductor nearest to the proposed barrier and, if mobile cranes are to work in the vicinity, the minimum distance between the barrier and the lines should be increased to 6 m plus the length of the jib.

The barriers may be either stout fencing, ballasted oil drums, timber baulks or an earth bank of at least 1 m height. Any fences or drums etc. should be painted a distinctive colour to aid visibility. An additional safeguard is to erect a line of high visibility bunting at a height of 3 to 6 m immediately above the ground level barrier, but it must be emphasized that this is only a secondary safeguard and must not be used as the only protection (see Fig. 18.1).

Passage of plant

In cases where plant must pass under overhead lines the first consideration is to restrict the width of this passageway to a minimum. The passageway underneath the lines should be fenced at ground level and provided at each end with 'goal posts' to act as gateways. The crossbars on the goal posts should be set at a height previously

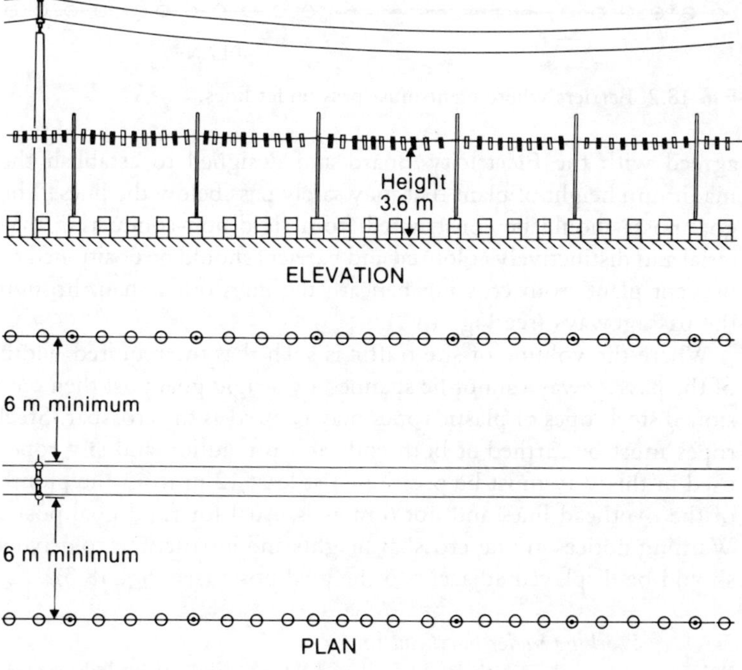

Fig. 18.1 Sites where there will be no work or passage of plant under lines.

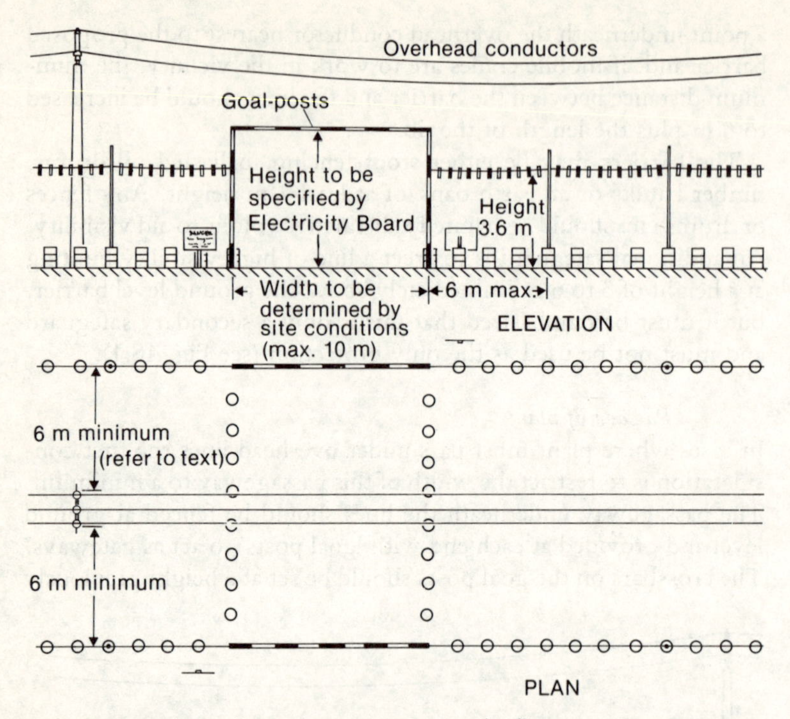

Fig. 18.2 Barriers where plant must pass under lines.

agreed with the Electricity Board and designed to establish the maximum height of plant that may safely pass below the lines. The goal posts should be constructed from rigid non-conducting material and distinctively coloured and barriers should be positioned to prevent plant from crossing beneath the lines other than through the passageways (see Fig. 18.2).

Where the volume of site traffic is such that the required width of the passageway cannot be spanned by a rigid goal post then tensioned steel ropes or plastic ropes may be used as the crossbar. Steel ropes must be earthed at both ends as a precaution and any ropes used in this way must be positioned at least 12 m from the plumb of the overhead lines and not 6 m as is usual for rigid goal posts. Warning notices giving crossbar heights and instructions to drivers should be displayed adjacent to the goal posts (see Fig. 18.3).

Working under overhead lines

When work underneath live overhead lines is unavoidable barriers, goal posts, etc., should be provided as necessary to limit the work

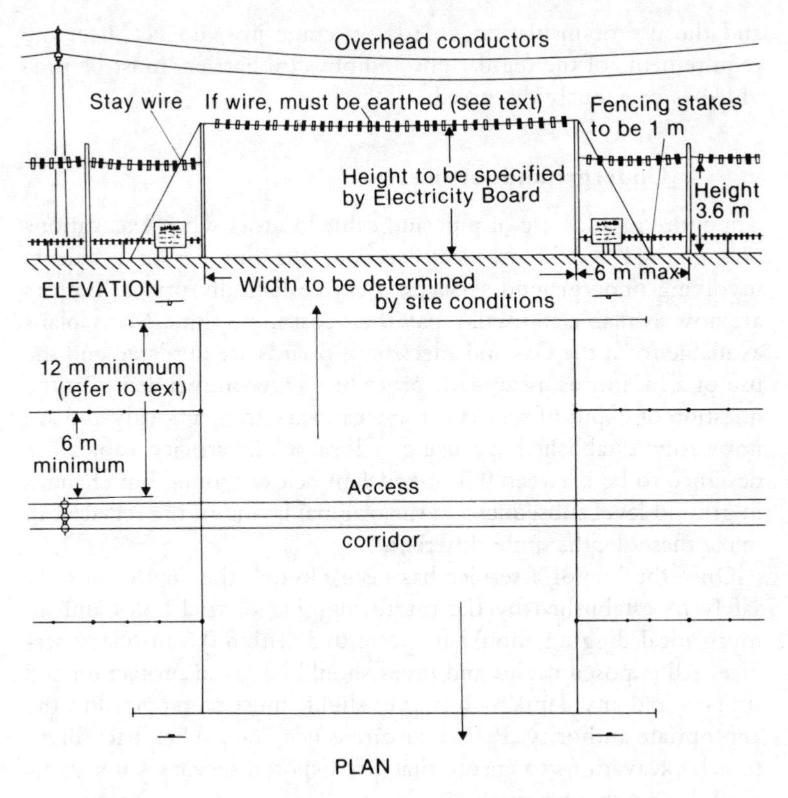

Fig. 18.3 Use of tensioned ropes when corridor is wide.

area to a minimum but obviously additional precautions become necessary.

The Electricity Board must be approached and a safe working clearance agreed and then a careful planning exercise is necessary to ensure that this clearance cannot be violated by plant and transport. Cranes and excavators can and must be fitted with physical restraints to prevent derricking and slewing movements likely to run into the danger area. Great care must be taken to ensure that the headroom is not reduced during the work operation by a build-up of the ground level and an 'overkill' of supervision is most desirable.

A word on warning devices fitted on crane jibs: they are a useful warning device but in most cases do only give a warning similar to that given by a load indicator when the safe load is exceeded. This warning may be ignored by the driver. Devices of this sort, and the use of insulating guards on crane jibs, do *not* meet the requirements of the regulations and physical barriers must be provided as previously discussed.

Underground services

The more regular use of pipe and cable locators when excavations are required would drastically reduce the number of accidents involving underground services. Very realistically priced locators are now available and will repay their cost in no time. Many plans available from the Gas and Electricity Boards are outdated and the use of a locator as a standard procedure is recommended. On the question of depth of services, these can vary tremendously and are not easily established by using a locator. Electricity cables are designed to be between 0.5 m and 1 m below ground but changes in ground level subsequent to the original laying of the service can make these depths quite different.

Once the line of a service has been plotted, the depth can only safely be established by the careful digging of trial holes and no mechanical digging should be permitted within 0.5 m of the service. All exposed cables and pipes should be given protection and support and any damage, however slight, must be reported to the appropriate authority. Particular care is necessary when backfilling trench excavations to ensure that any exposed service is not damaged during the process.

Fire and explosion risks

When excavations are being executed near to gas mains regular checks should be made to make sure no gas is escaping. Town gas and natural gas are both lighter than air so are more easily dissipated but the risk still remains and no chances should be taken. Damage to or working on live sewer lines can also present risks of explosion due to the possibility of flammable gases being present and additionally the dangers of toxic gases must also be considered.

19

Refurbishing work

Lack of suitable space for new building, preservation orders on older buildings, escalating costs of new building; all these factors have contributed to the growth of the refurbishing industry within construction. What the home owner has been doing in a modest way for years is now relatively big business in the commercial and industrial field.

Refurbishing also brings new problems for the builder, probably not major ones, but certainly problems that can have serious health and safety consequences. All types of buildings are refurbished – offices, public buildings, stores and most of all housing. Sometimes it is housing which is still occupied, which adds yet again to the problems, and it is this class of refurbishing work upon which we will concentrate, believing that most problems common to all types of refurbishing will be covered.

Planning the work

One basic important rule must always be followed. *Never assume that the fabric of the building is sound.* It is useless erecting a perfect scaffold and tying it into a brick wall that is about to collapse; similarly, fixing a temporary guard rail to an existing timber staircase whose treads are poorly supported is somewhat pointless.

The main problem is that the extent of repairs necessary is difficult to establish until work commences. The fabric of the building is usually completely covered with plaster etc. and a careful investigatory survey is an essential first step.

Many buildings which are being refurbished are from 30 to 130 years old and the methods of construction acceptable in those times were vastly different from the building regulations of today. Timber was never treated against rot or beetle attack, timber lintels were used over door and window openings and the joists were built into the brickwork without any protection. All these areas should be considered as possible hazards and the removal of window and door frames and floor stripping operations must be undertaken with extreme caution.

Scaffolding and access

Scaffolding for this type of contract is similar to that provided for a demolition contract. The main structure is already standing so the normal procedure is to erect the scaffolding to the full height of the building, attend to any necessary roof repairs or alterations and then deal with external decoration, plumbing etc. from the top downwards.

On contracts where major roof works are required it may be necessary to install a temporary roof. This is a skilled scaffolding operation and must be undertaken by competent scaffolders. The additional loading of the roof cover and wind loads against the side sheeting makes the tying in of the scaffold vitally important and emphasizes again the need for careful inspection of that part of the structure to which the scaffold is to be tied.

Temporary works

It is generally necessary to do a certain amount of cutting of new openings when altering and refurbishing a building and this is another possible hazard area which needs careful attention.

If any main load bearing walls are to be removed or radically altered there will obviously be a need for falsework support. Unlike most falsework in new construction the problem with applying support in the refurbishing situation is that there is often difficulty in transferring the load to solid ground. It is important that this is achieved, however, and you may find it necessary to seek expert advice on this point from your company or any other design engineer.

Rot and timber infestation

Another item which can lead to major structural problems is timber infestation. A great deal of timber is used as structural members in older type buildings and can be affected by dry rot, wet rot, beetle and fungus attack.

Whenever rot or beetle attack is suspected temporary support must be erected immediately and any infected timber that is removed should be burned on site or, if this is not permitted, should be transported from site in covered containers.

Beetle infested timber can often be treated successfully, depending on the extent of the damage, and a word of warning is necessary on this subject. The treatment is to spray with a material which both kills existing beetles and also leaves a toxic residue for the benefit of any future beetles.

The spray is often highly flammable and toxic to human beings and, whilst the specialist firms who carry out this type of work protect their own operatives, it is important that no other operative accidentally enters the area being treated.

Suitable warning notices should be displayed and good ventilation provided after the spraying has been completed, and the area closed from access for the required period.

Asbestos

The use of and working with asbestos is now carefully controlled and Chapter 15 deals with this subject in some detail, but hundreds of tonnes of asbestos of various types are present in old buildings in the form of lagging of boilers etc.

If any evidence of asbestos is found, work must stop, a sample of the material must be tested and appropriate precautions taken.

Statutory services

On most refurbishing contracts all mains services will probably have been disconnected but this is certainly not necessarily the case when only *parts* of buildings are being dealt with. What is important is that no assumptions of dead electrics or gas mains are made in *any* old building. Regularly cases are reported of services in supposedly long empty dwellings being found by refurbishing contractors to be live and operational. Many such cases are assumed to have resulted from the activities of squatters in the premises and there is no doubt that this section of the community can take credit for many of the ingenious connections but, whoever is responsible, the supervisor must ensure that a check-out of the services receives top priority when taking occupation. Provision of adequate temporary facilities will, of course, be necessary but these must be properly installed, as for a new construction contract, and any old existing services disregarded and made dead.

Work in occupied premises

There are several problems which should be borne in mind when dealing with this particular type of contract. First there is the need to maintain good and safe access and escape both for workmen and also for the residents who are occupying part of the building. It is important that solid divisions are erected to prevent members of the public straying into the building area, but any divisions so provided must not infringe upon access and escape routes. If this is unavoidable, temporary safe routes must be provided and clearly indicated.

Maintenance of services is another area that needs careful planning; gas, water, electricity and drainage all can create problems and safety hazards if not controlled.

The control of dust can be a major problem. Many years' accumulation of dust will have settled behind plaster and under floorboards and when disturbed much will become airborne. The dust should not only be cleaned up regularly, using vacuum equipment, to ensure a reasonable working area for the work-force, but the division screens should be adequate to safeguard the residents and the general public from the airborne dust.

Not many operations in the construction industry can be said to be quiet and noise and vibration when working in occupied premises can present difficulties. Much can be done to reduce these problems but again 'planning' is the key word. The use of 'quiet' compressors, electrically operated hoists and mixers can all make useful reductions in the noise and vibration levels but need to be planned for and ordered on time.

Possibly the biggest problem is that of security. This has been touched on in the previous reference to divisions between the working area and any occupied area but needs to be emphasized.

Work areas *must* be left secure when the work finishes for the day and this not only means locks, bolts and bars to prevent trespass but also means leaving the work area *within* the compound safe. Contractors have a very real responsibility towards children and if a child manages to beat your security fencing he, or she, although a trespasser, must not be exposed to unnecessary risks.

20

Lasers in construction

The word 'laser' is an acronym of *l*ight *a*mplification by *s*timulated *e*mission of *r*adiation and is, in fact, a device that produces a beam of light of very high energy. Since the beam is straight and does not spread like the beam from a torch, it is used widely on construction sites for surveying and levelling operations. Lasers are also used in the medical field for microsurgery and delicate operations such as sealing off tiny blood vessels and it is this use which gives us the clue that there could be some dangers from the instruments and the beam.

Hazards

The beam from a laser may cause injury to both the eye and the skin but it is particularly the eye which is at risk. In the event of a laser beam entering the eye the lens will focus the beam to a tiny spot on the retina and this focussing effect amplifies the energy density and causes a rapid rise in temperature and consequent burning. A burn may also occur when the skin is exposed to laser radiation but generally this would require a beam of fairly high energy and more lengthy exposure. Damage can also be caused if the beam is reflected into the eye. Polished surfaces are obviously the most dangerous but light colours can produce reflection of sufficient danger.

It is therefore clear that the laser used on our sites as a surveying instrument is worthy of respect. The hazards can largely be eliminated by ensuring that makers' instructions for the user are strictly complied with but, as in many things, familiarity can breed contempt and only strong supervision will prevent risks from being taken.

Safety measures

1. The first and most important rule is to insist that only trained employees are permitted to set up and use laser equipment and that they fully understand the responsibility *they* have for the safety of other operatives, visitors and 'passers-by'.

   This training, which may be given by the manufacturer or supplier, by a safety officer qualified in the subject or by an approved training organization, should include:
   (a) safe operating procedures;
   (b) the proper use of hazard control procedures, warning signs, etc;
   (c) the need for personal protection;
   (d) accident reporting procedures.
2. All areas in which lasers are used should be posted with a standard laser warning sign (see Fig. 20.1).

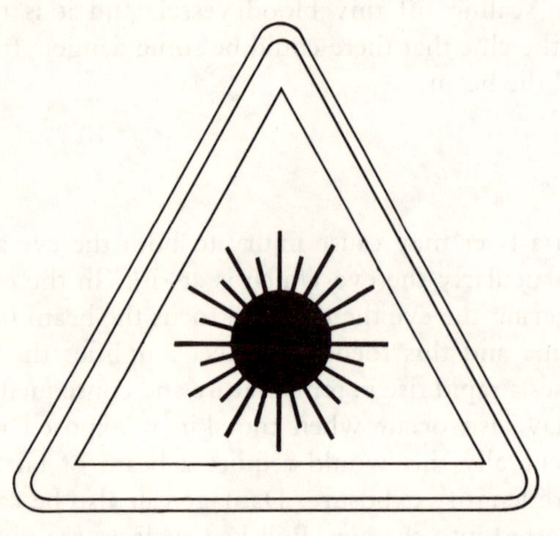

Fig. 20.1 Caution – laser beam. Symbol and border – black; background – yellow.

3. Great care should be taken when aligning the laser. Wherever practicable, mechanical or electronic means should be used, as direct viewing along the beam can be risky should any reflection take place.
4. All equipment must be labelled to indicate the maximum output power. Those used in construction will be less than 5 mW (see Fig. 20.2).

Fig. 20.2 Warning label. Label wording and border – black; background – red.

5. The beam must be efficiently terminated at the end of its path and never permitted to travel beyond site area.
6. Precautions should be taken to make sure that persons do not look directly into the beam or use optical instruments to view the beam and to help in this matter the beam path should, wherever practicable, be located well above or below eye level.
7. Reflective materials and objects should be removed from the beam path or adequately covered.
8. The inadvertent or deliberate tracking of vehicles and aircraft must be prohibited. Cases have been reported of drivers and pilots being affected by badly directed beams with very serious results and any tendency towards skylarking of this sort must be dealt with severely.
9. When the lasers are not in use they should be locked off by the user and stored where unauthorized personnel cannot gain access.

There is no doubt that the laser instrument is a very useful aid for the surveyors and site engineering staff, particularly on extensive pipeline projects, and that when properly used there is no danger but, unless due precautions are taken, severe injuries to the eye can occur and the site supervisor should insist that the measures listed above are strictly adhered to.

21

Explosives

The quick and efficient excavation of rock is a vital operation in civil engineering work and the careful and expert use of explosives can be an important aid in this operation.

In the same way, and as has already been discussed in Chapter 5, explosives have their uses in demolition work, again provided that expert and experienced men are carrying out the work.

I have tried to emphasize the need for training throughout this book but in the case of explosives the necessity for using only trained men is imperative.

This is *really* a job where the site supervisor should either call in a specialist firm or recruit an experienced shot firer, and accordingly I shall only discuss fairly basic operational techniques and concentrate in this chapter on the more general aspects of storage and transport of explosives.

Obtaining explosives

Understandably, the acquisition of explosives is rigidly controlled and, except in the case of Government Departments and holders of licensed storage magazines, all prospective purchasers must have, depending on which category of explosive they require, either a police certificate or a police licence.

A 'licence' only permits purchase of blasting powder (gunpowder) and safety fuse, and a 'certificate' is necessary for the purchase of all other types of high explosives, detonators and igniter cords. It is from this second range of explosives, those for which a purchase 'certificate' is required, that materials will be chosen for use on a construction site and, for all practical purposes, blasting powder may be disregarded.

Licences are granted by the Chief Officer of Police who will satisfy himself that the person or firm to whom the licence is granted will take every precaution necessary to ensure public safety. Purchase licences are normally valid for a period of twelve months and will usually be issued to run for the same period as the stores licence.

Explosives for immediate use

When explosives are required for a one-off job the Chief of Police may issue an 'immediate use' certificate which will cover just the one transaction.

Technically, explosives purchased in this way should be completely used on the day of receipt but the authorities recognize that this is not always possible and generally raise no objection, where delay *must* occur, to explosives being stored for up to three or four days. In such cases the Police and the Explosives Officer appointed by the local authority should be advised and will need to be satisfied as to the measures proposed to ensure the security and safe storage of the explosives. If a delay of longer than three or four days is likely the local authority may consider it necessary to require the storage place to be registered.

Storage of explosives

There are several different sets of requirements for storing explosives, based sensibly on the amount to be stored.

Private use

For amounts up to 4.5 kg of explosives and 100 detonators no legal restrictions apply, but it is strongly recommended that the requirements as for Mode B–Registered Premises are followed. This requires that the explosives should be kept in a stout wooden, or wooden lined, box and that the box should be provided with a good lock. Detonators *must* be stored separately , again in a lockable wooden box, and both boxes should be clearly labelled 'explosives' and kept in a secure building, preferably near an exit.

Mode B–Registered Premises

The storage of mixed explosives up to 7 kg in weight comes within the scope of the Explosives Acts and Regulations and has already been described in the previous paragraph.

Mode A–Registered Premises

Up to 27 kg of mixed explosives may be kept in Mode A registered premises.

This requires a completely detached building constructed of brick, stone, concrete or steel which must be at least 15 m from any public thoroughfare or place where people congregate or work.

Registered premises must be kept scrupulously clean and free from grit and used solely for the storage of explosives. Any tools used in connection with the premises should be of copper, bronze or other non-spark metal. It must be emphasized again that detonators MUST be kept in a separate annexe.

A design suggested by ICI for Mode A Registered Premises is reproduced in Fig. 21.1 and illustrates the importance which is placed on safe and secure storage. All registered premises must be registered and approved by the local authority.

Explosives stores

For quantities in excess of 27 kg a properly constructed building is required, which again must be licensed by the local authority.

Stores are categorized as follows:

Division A – weight of explosives not to exceed 68 kg
Division B – weight of explosives not to exceed 136 kg
Division C – weight of explosives not to exceed 453 kg
Division D – weight of explosives not to exceed 906 kg
Division E – weight of explosives not to exceed 1812 kg

and the separation distances between the stores and other property increases as the amount of explosives stored rises, as indicated in Table 21.1

A design suggested by ICI for a mixed explosive store is reproduced in Fig. 21.2 on p. 163.

Lightning conductors

These must be provided for all stores other than Divisions A and B, that is those containing in excess of 136 kg of mixed explosives.

Electric lighting

In larger stores, where electric lighting is necessary, strict regulations apply. These are dealt with fully in the pamphlet 'Electric Light and Power in Factories and Magazines for Explosives' pub-

Table 21.1 Property separation

Class 1 Property		Class 2 Property	
Dwelling house. Shop. Room of any kind. Workshop. Railway. Furnace or kiln. Magazine for explosive. Store for explosive. Registered premises.	In occupation of licensee or with consent in writing of occupier.	Dwelling house. Shop. Room of any kind. Workshop. Railway. Furnace or kiln. Magazine for explosive. Store for explosive. Registered premises.	Without consent of occupier.
Highway, footpath or public open place or any work place. Canal or navigable water. Dock, river wall or sea wall. Pier or jetty. Reservoir.		Factory. Building in care of D of E. Place of public worship. Educational establishment. Hospital or like institution. Court of Justice. Theatre, cinema, covered market or other covered building where the public are accustomed to assemble. Government or local government building.	

Store category	Distance required from Class 1 Property, m	Distance required from Class 2 Property, m
Division A	23	26
Division B	23	40
Division C	45	90
Division D	70	140
Division E	107	214

lished by and obtainable from the Home Office, Explosives Branch. The following notes are provided for general guidance.

Where electric lighting is required, dustproof fittings with machined metal-to-metal joints should be used. For pendant fittings, conduits should be laid in the roof concrete to each fitting with a minimum of conduit drop inside the building. Each fitting

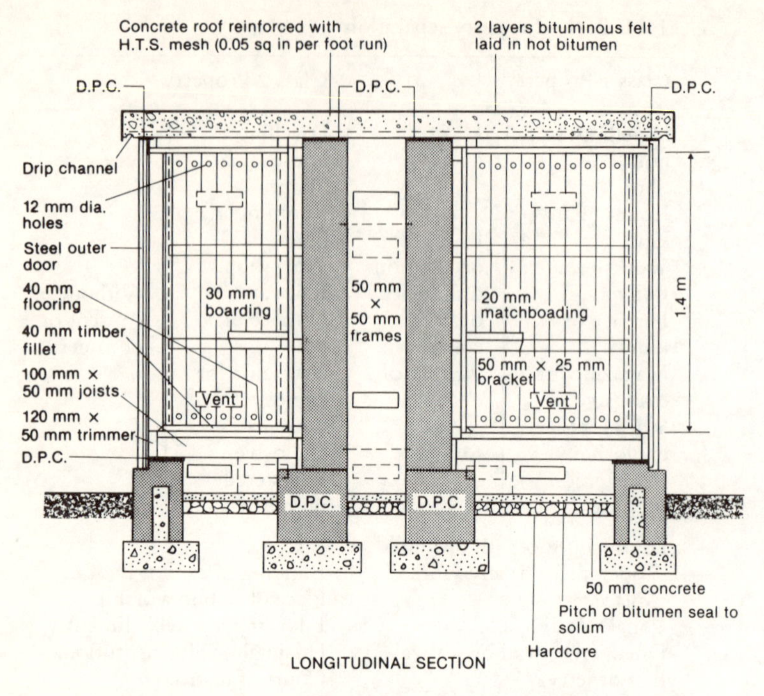

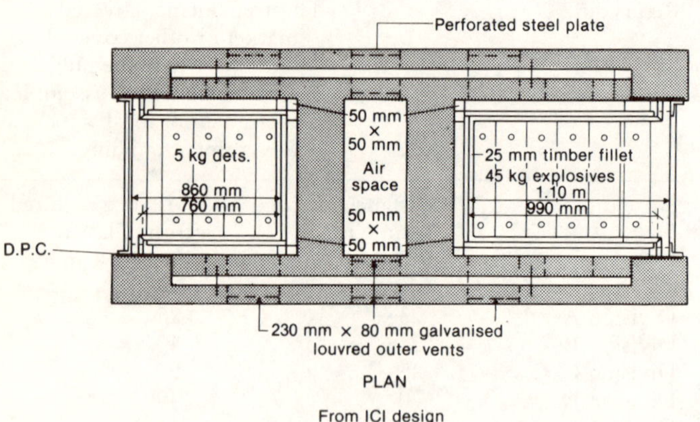

Fig. 21.1 Registered premises – Mode A.

should be independently supported from the ceiling by two galvanized suspension wires. No junction boxes or conduits are allowed inside the building and any junction boxes used on the outside walls should be weatherproof.

All wiring should be carried out in 3/0.91 mm. CMA grade PVC

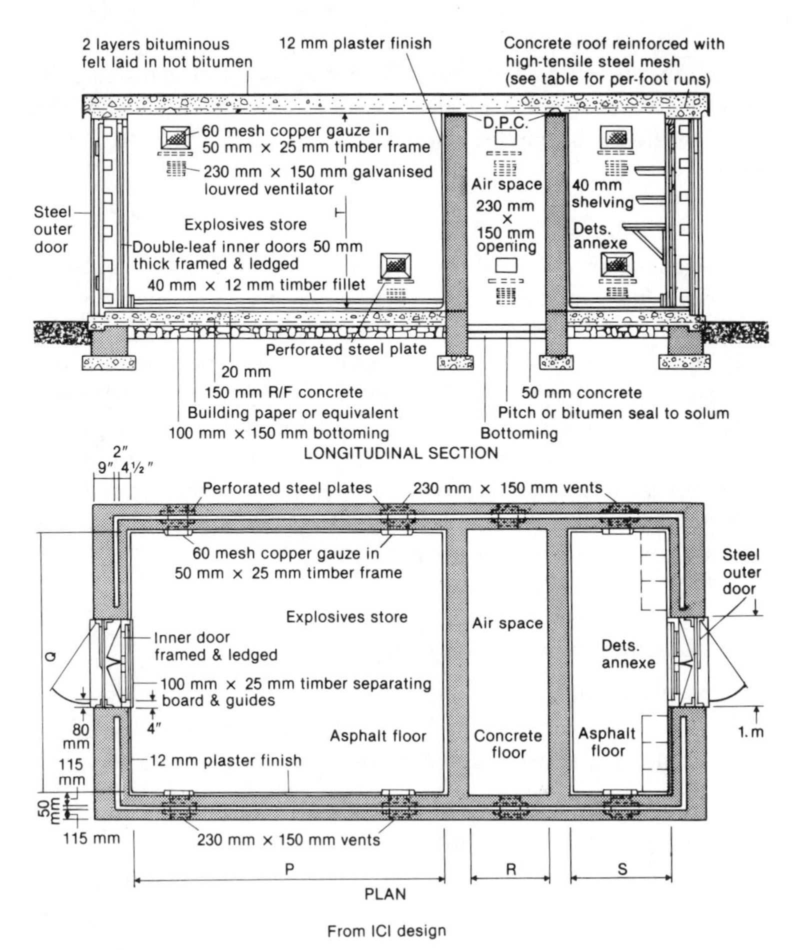

Fig. 21.2 Store for mixed explosives.

cable enclosed in 19 mm diameter heavy gauge solid drawn galvanised conduit.

The switching of lights must be carried out from a position outside the building. The usual practice is to employ non-certified switchgear at a distance of not less than 8 m and preferably 20 m from the building. All switches should be double-pole and the supply to each building must be controlled by a master switch capable of isolating every conductor entering the building.

Where buildings are supplied with electricity by overhead lines, the overhead system must not approach within 15 m of the building, the remaining distance being completed by underground cable.

Arrangement of explosive cases

Cases must be stacked at least 150 mm away from walls so as to provide free movement of air. The oldest stocks should be used first and the cases must be stacked in accordance with instructions on them.

Repairs

If any repairs are required to the store, all the explosives must be removed and care taken also to remove any explosive dust by washing out the interior.

If the repairs can all be done between sunrise and sunset of one day usually no objection will be raised to removing the explosives onto planks in the open and covering them over with tarpaulin. Should the repairs be of longer duration an arrangement must be made for putting the explosives into some other licensed place of storage. Before repairs are put in hand the stock in the store should be run down to its lowest point.

Notices

The following notices must be displayed inside stores:

1. A copy of the general rules appertaining to the store.
2. An extract from the licence showing the quantity of explosives which may be lawfully kept in the store.

In addition, a 'Warning to Trespassers' notice must be prominently displayed outside the building.

The advice of the police should be sought on the wording of warning notices as they are understandably touchy about notices which can publicise the whereabouts of explosives.

Transporting explosives

Up to 45 kg of explosives may be transported by private vehicles, not specially constructed for the job of transporting explosives, provided that all due precautions are taken to prevent the possibility of accidents.

Explosives should be carried in their original packings inside strong metal containers and preferably be sited on the floor of the rear seat compartment. The driver, and any passenger, should refrain from smoking and the vehicle must never be left unattended.

Larger quantities of explosives, up to 230 kg, may be carried in motor vehicles that have not been specially constructed provided certain additional safeguards are met:

1. They should be contained in approved metal cases and rest on the floor of the vehicle.
2. If the vehicle has an open body and the sides and back of the body are less than 500 mm high, the load must be tied down by a tarpaulin.
3. Two people must always accompany the vehicle.
4. An efficient chemical fire extinguisher must be carried with the vehicle and no other chemicals, oil drums, petrol or other flammable material must be carried.

Quantities of explosives in excess of 230 kg must be carried in vehicles specially constructed for the purpose but transportations of this size will normally be undertaken by manufacturers' vehicles.

Use of explosives

Whenever it is decided to make use of explosives on a contract site the site supervisor should, in conjunction with the specialist sub-contractor or his own shot firer, draw up a set of rules to ensure the safety of the site.

This set of rules should include:

1. A system of notices warning anyone who may approach the danger zone.
2. An internal warning system designed to alert all operatives.
3. The provision of a distinctive 'alert' and 'all clear' signal for each blasting operation.
4. The posting of sentries at the boundaries of the danger zone both to inform the shot firer that the area is clear and also to prevent entry into the danger area until the 'all clear' has been sounded.

The shot firer

Shot firers must be experienced men, not less than 21 years of age, and should be appointed by the site supervisor in writing. The supervisor should ensure that the shot firer is aware of the responsibilities and that he works to the agreed rules described above.

Procedure in charging shot holes

Charging shot holes is the responsibility of the shot firer and there are several safety rules that he must follow and of which the site supervisor should be aware:

1. Accurate drilling and charging is all-important in excavation work and holes must be drilled to the pattern decided and to an adequate size and depth. As secondary blasting is difficult and expensive every care should be taken to ensure that the rock is broken sufficiently at the first blast and it is usually necessary to drill to between 300 mm and 1 m deeper than the excavation depth required. Drill holes should always be at least 3 mm larger in diameter than the cartridge diameter to be used.
2. The shot hole should be checked for depth and clearance.
3. For most operations electric detonators are to be preferred because of the ease with which remote control can be exercised. When using this system a primer cartridge must first be prepared by making a hole at one end of an explosive cartridge with an aluminium, brass or wooden pricker and pressing the detonator into the hole until it is completely buried in the cartridge. The wires leading to the detonator should then be secured around the cartridge to prevent any possibility of them being fractured or the detonator being disturbed when the cartridge is placed in the hole. This primer cartridge is then inserted into the hole and the remaining cartridges, to make up the agreed charge, inserted one at a time on top, taking care not to damage the covering of the wires leading to the detonator.
4. The shot firer must NEVER remove the wrapping paper from an explosive cartridge, NEVER use metal rods for charging and stemming holes and NEVER use force in pushing cartridges home.
5. When all the cartridges have been placed in the hole some 0.609 m (2 ft) of stemming or filling should be gently tamped on top, again taking care not to damage or strip the detonator wires.
6. This procedure is repeated until all holes in the blast have been charged.

Firing of shots

It is then the duty of the shot firer to make the final connections.

He does this by linking the detonator leading wires together in series and making connections to the firing cable.

When this has been completed, but BEFORE the 'exploder' is attached, the shot firer must make sure that all the safety precautions for personnel previously discussed are put into effect and, when completely satisfied, he should connect the exploder, insert the firing key and fire the blast.

Immediately after firing, the shot firer should remove the firing key from the exploder and retain it in his personal possession and disconnect the exploder from the cable.

Finally, he should wait for 10 minutes and then examine the area for misfires before sounding the 'all clear'.

General precautions

A few additional precautions of a general nature should be mentioned:

1. When drilling shot holes never permit re-drilling of old holes as an explosive charge may be present which could be detonated by the drill.
2. Prohibit smoking within 6 m of any explosive, detonator or the appropriate store.
3. Allow a minimum amount of explosives to be drawn from stores at a time.
4. Ensure that, when shots are fired using electric detonators, the shot firing cable is of sufficient length to enable the shot firer to operate from a safe place.
5. Be particularly careful not to use electrical detonation methods during electrical storms and also in the following circumstances:
   (a) within 20 m of 11–70 kV
   and 60 m of 132–400 kV overhead electric cables;
   (b) within 1.5 km of a radio or TV station.
6. In the same context 'walkie-talkie' communication systems should not be used within 100 m of any electrical detonation blast.
7. If there is any possibility of the blasting operation affecting the public in any way then the police should be notified and their advice sought.

22

Radiation hazards

The use of radiography as a means of 'non-destructive testing' of metal pipework, boilers, etc., is now quite common in the construction industry. In many instances the external surfaces of a metal component will appear to be perfect and, before radiography came into use, the only way effectively to check for flaws was to cut the metal. The use of an X-ray machine or a radioactive source has the advantage that there is no damage to the component during the testing process and the method is specially valuable for the testing of welds.

It does, however, have the drawback that, unless proper precautions are taken, this form of radiation known as 'ionizing radiation' can injure the cells of the body. It can cause many illnesses, including cancer, and in large doses can even cause death.

Such a specialist piece of equipment will naturally only appear on your site either when under the care of an experienced radiographer employed by your company or when operated by a subcontractor engaged to carry out special tests, and much of the relevant legislation will be the responsibility of these 'experts' or their superiors. The overall responsibility for the safety of the site still remains on the shoulders of the site supervisor and accordingly he should be aware of the main hazards and the controls that are necessary. The information that follows is of a general nature only and supervisors requiring further information should refer to the *Code of Practice for Site Radiography* published by Kluwer Publishing of London.

Sealed sources – gamma ray

The forms of ionizing radiation likely to be found in the construction industry are X-rays and gamma rays but, because of the portability and relative low cost, the most common equipment uses gamma ray sources and I shall concentrate on the problems that are associated with that type.

These isotopes comprise a radioactive source housed in a protective capsule. When in use the source is exposed and, when the operation has been completed, the source is retracted into its container. Unlike an X-ray machine the radioactive source cannot be switched off and it emits gamma rays spontaneously and continuously. The only control is to shield the source in its properly constructed housing. Another problem is that the radiation beams leave the source naturally in all directions and so obviously the housing needs to be constructed so that the angle of the radiation beam actually leaving the housing is limited (see Fig. 22.1).

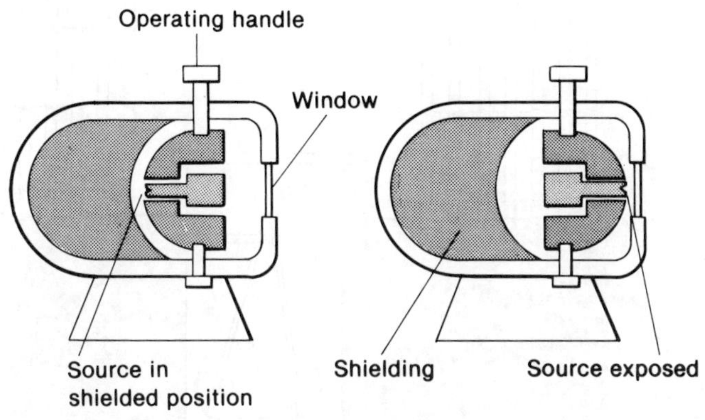

Fig. 22.1 Example of a shutter-type source container.

When in use the sources direct their rays through the object being examined and produce an interior photograph of the metal in much the same way as X-ray films of the human interior are produced by medical radiographers. The gamma rays give pictures that are actually shadowgraphs and any faults, in the form of blowholes or cracks, will show up as darker areas on the film.

Storage of isotopes

When the use of isotopes on your site is limited to either a 'one-off' visit or occasional infrequent visits, any storing on site becomes unnecessary as the equipment will normally be taken to and from the site daily. When, however, a considerable amount of radiography is to take place it may become necessary to store isotopes on site.

When this is the case you must ensure that specially-constructed storage pits are provided. The recommended method is to take a steel tube of about 450 mm to 600 mm diameter and about 1.5 m long with a welded base and sink it into the ground, leaving a protrusion sufficient to enable a lockable hinged lid to be fitted. A hanging bracket should be fixed near the top of the tube and the isotope suspended by rope from this bracket at a depth of about 1 m. The pit should be sited away from any vehicle or pedestrian traffic way and surrounded by a fence with a lockable gate. The fence should be at least 2 m from the pit and should bear warning notices (see Fig. 22.2).

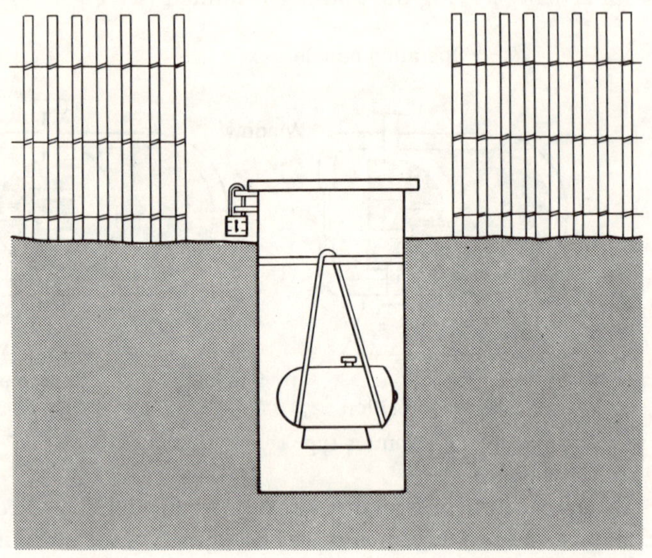

Fig. 22.2 Example of isotope storage pit. The pit to be completely fenced and warning notices displayed.

Using isotopes

The persons most at risk are the operators of the equipment, but it is not proposed to deal with this point in any detail as we have already assumed that the operators are specialist visitors to your site. You should satisfy yourself that they have been authorized and that they carry written evidence to that effect but other than that your concern should be for the safety of other site workers and visitors.

Working area

All persons except 'authorized persons' must be excluded from the work area. Inspection work on site will normally have to be done *in situ* and so it will be necessary to provide a suitably marked area, indicated by rope or bunting, set at an adequate distance from the radioactive source.

This adequate distance depends on the type of radioactive material used and the strength of that source, known as its 'activity'. The most common sources used in non-destructive testing radiography are iridium 192 and cobalt 60 and, just as an example of safe distances, a source of 10 curies of iridium demands a distance of approximately 80 m and the same strength source of cobalt requires approximately 125 m.

The curie is the unit used to denote the quantity of radioactive material in the source. These are approximations and the only sure method of determining the safe distance is by using a dose rate meter around the barrier boundary. The reading on the meter should not exceed 0.75 millirems per hour which it is not necessary to explain other than to say that this is approximately the reading you will obtain at 80 m distance from 10 curies of iridium in the example quoted above.

Notices should be displayed prominently around the barrier bearing the words 'Do not pass this barrier – radiography in progress' and showing the hazard pictogram for radiation (see Fig. 22.3).

Fig. 22.3 Safety sign. Caution: risk of ionizing radiation. Symbol and border – black; background – yellow.

Warning signals

Whenever a sealed source is *about* to be exposed a warning signal should be given to all persons in the vicinity and a second different signal should be given and maintained during the time that it *is* exposed. Warning lamps are generally used for this purpose and the supervisor should make sure that the lamps are positioned so that they are clearly visible and that all site personnel understand their meaning.

Measurements of dose rate

Regular measurements should be made with a dose rate meter around the boundary of the barrier to ensure that the rate does not exceed the set 0.75 millirems per hour. This will normally be undertaken by the radiographers but it is in the interests of all concerned that the site supervisor ensures that the checking is being carried out.

Effects of radiation

The effects of a dose of ionizing radiation vary according to the type of exposure; whether the exposure is local, affecting only a part of the body; or general, affecting the whole body. In addition, the time that the body is exposed to the source also affects the seriousness of the consequences.

Local exposure usually results in a reddening or blotching of the skin and in severe cases ulcers may be formed, but the effects of acute *general* exposure range from mild sickness to a severe illness with vomiting, diarrhoea, collapse and death. General exposure to small doses over a long time can result in anaemia and leukaemia.

In short *any* radiation exposure must be considered to be a serious matter and treated as potentially damaging, which emphasizes the importance of ensuring that only 'authorized persons', who are experienced in these matters, enter the hazard areas. This is where the site supervisor can and must play his part. His instructions about the restricted areas and warning signals must be absolutely clear and he must make quite sure that they are obeyed.

23

Permit to work systems

There are certain operations that take place on contract sites where the safety of the operatives directly concerned and of other workers in the area depends more than usual on a clear understanding of instructions.

Maintenance work on plant and equipment or on electrical or liquefied petroleum gas supplies; entry into confined spaces and live sewers; welding in possibly dangerous atmospheres; all operations such as these, whether done as part of a planned maintenance system or in an emergency situation, always necessitate some departure from ordinary routine.

Unless the operation is carefully planned beforehand and competently supervised dangerous situations can develop in which men may be put at risk either by the actions of others or by their own failure to realise that they are in a position of actual or potential danger. Occasionally, safe working conditions may be achieved by the use of automatic safeguards but more often reliance has to be placed upon actions or reactions of people. Verbal instructions, requests and promises are always liable to be misheard, misinterpreted or forgotten and should never be regarded as a satisfactory basis for action on which men's lives may depend.

To achieve the maximum degree of safety the human element must be eliminated as far as possible and the use of a 'permit to work' system, which requires that a written and signed instruction be in the hands of the man in charge of the operation before work is begun, is to be strongly recommended.

The permit – general principles

In operating a 'permit to work' system the following principles should be observed:

1. The information given in the permit must be detailed and accurate. It must state exactly what work is to be done and must specify the steps that have been taken to make the plant or operation safe.
2. The permit must specify the time that it comes into effect and should be recognized as the master instruction which, until it is cancelled, overrides all other instructions.
3. No one must undertake any work whatsoever that is not specified in the permit, or work in an area not specified as being safe. If it is found that a programme of work *must* be changed, no variation should be introduced until after the existing permit has been cancelled and a new one issued. The only person who has the authority to alter the programme and issue a new permit to work is the person who issued the original permit or whosoever is properly authorized to take over responsibility from him.
4. Anyone who does take over from the person who issued a permit must assume full responsibility for the permit until it is either cancelled or handed over to another authorized person.
5. Any person issuing a permit must, before signing it, make sure that all the specified precautions have in fact been taken and should, unless absolutely impracticable, check this by personal inspection.
6. The person who accepts the permit becomes, from that moment, responsible for ensuring that all the specified safety precautions are met, that only permitted work is done and that this is confined to the work area defined in the permit.
7. It is good practice to exhibit a copy of the permit on site whilst it remains in force, as a reminder to all concerned, and it is particularly important that all parties who are likely to be involved are advised, consulted and instructed as appropriate.

It is not suggested that the 'permit to work' system can entirely eliminate accidents but a properly controlled and effective system can play a substantial part in reducing the part played by human failures. The site supervisor is advised to consider following the system whenever faced with particularly hazardous work situations of the types previously discussed.

Part D

Health and the environment

24

Health and welfare

The facilities provided for health and welfare on construction sites are all too often the 'cinderella' of the organization, particularly on the smaller site. Bad planning is the main culprit. Delivery arrangements that provide for, and are geared to, early production often lead to work being started on the main project before adequate welfare facilities have been provided. The wise supervisor will guard against this and ensure that a first priority on taking possession of a site is the setting up of the necessary, and legally required, facilities.

Welfare facilities

The requirements for welfare facilities on site are dealt with in a special code of regulations, the Construction Health and Welfare Regulations 1966, which sets out various levels of requirement, depending on the number of persons employed, for each different facility.

These have been drawn together to show the total welfare package legally required for a given labour force, but these should be looked upon as minimum requirements and the good employer will obviously try to do better than the absolute minimum. He may also find it administratively easier to adopt say the 'more than twenty men on site' standard for all contracts and build from there for larger sites.

On every site

1. At least one sanitary convenience must be provided on every site. The regulation requires these to be provided at the rate of one for every twenty-five persons on site.

Sanitary conveniences must be reasonably accessible, be under cover, ventilated, have a door with a fastening and be provided with lighting. They must be kept clean and not open directly into mess rooms or work rooms. Separate conveniences must be provided for men and women.

2. Shelter must be provided for protection during bad weather; for storing personal clothing; for storing protective clothing; and for taking meals, with seating accommodation and facilities for boiling water.
3. Drinking water must be provided at convenient points on every site and clearly marked 'drinking water'.
4. Protective clothing must be provided for all operatives who are required to continue working in rain, snow, sleet or hail.
5. Washing facilities must be provided on every site where anyone is employed for more than four hours and, if the site is expected to last for more than six weeks, these facilities must be at the standard required for a site employing more than 20 men.

More than 5 men on site

In addition to the facilities previously listed, sites employing more than 5 men must provide shelters that are heated for men and clothing and arrangements for drying clothes.

More than 10 men on site

Additionally there must be provided, in the mess room, facilities for heating food, unless hot food is available elsewhere on the site.

More than 20 men on site

On sites where a contractor employs more than 20 men, or where the work is expected to last more than six weeks, the minimum standards of the washing facilities are laid down and must include:

1. Troughs, basins and buckets.
2. Soap and towels or driers.
3. Hot and cold or warm, water.

More than 100 men on site

Where more than 100 men are employed, and the work is expected to last for more than twelve months, the regulations are more precise about washing facilities and specify that four wash basins are to be provided, plus an extra one for every 35 men in excess of 100.

First aid

The requirements for first aid at work are set out in the Health and Safety (First Aid) Regulations 1981 and the associated approved code of practice. Detailed study of these documents, issued as Health and Safety booklet HS(R)11 available from HMSO, is recommended but the main points concerning construction sites are indicated below.

Duty of employers – equipment

The employer must provide such equipment and facilities as are adequate and appropriate for enabling first aid to be rendered. The requirements for the contents of first aid kits vary according to the number of employees at the work location as shown in Table 24.1. Details of the requirements for 'travelling first aid kits' are also given. These kits should be issued where employees work alone or in small groups in isolated locations.

Where 250 or more employees are at work on a site a suitable first aid room should be provided. In addition, a sufficient number of first aid boxes should be provided for any work area that cannot be reached from the first aid room in approximately three minutes.

Duty of employers – first aiders

At least one trained first-aider should be present when the number of employees at work is between 50 and 150 with at least one additional first-aider for every 150 or so employees.

Where less than 50 employees are present then an 'appointed person' must be provided. The appointed person need not be a trained first-aider but it is strongly recommended that he has *at least* a knowledge of emergency first aid (see Chapter 27).

Where special or unusual hazards are present, the first-aider

Table 24.1 Required contents of first aid kits and 'travelling kits'.

Item	Numbers of employees			
	1–10	11–50	51–150	Travelling kit
Guidance card	1	1	1	1
Wrapped sterile adhesive dressings	20	40	40	6
Sterile eye pads	2	4	8	
Triangular bandages	2	4	8	1
Sterile covering for serious wounds	2	4	8	1
Safety pins	6	12	12	6
Medium sized sterile unmedicated dressings	6	8	12	1
Large sterile unmedicated dressings	2	4	10	
Extra large sterile unmedicated dressings	2	4	8	

Note: Where tap water is not readily available, sterile water or sterile normal saline in 300 ml disposable containers must be kept near the first aid box in the following quantities (according to the number of employees given above): 1, 3, 6.

should be an 'occupational first-aider' who has received approved training in first aid treatments relative to the particular hazards on site, in addition to the standard training. Many sites will fall within this category.

Other requirements

Several other regulations are of a general nature and should be mentioned.

Safe access

Safe access must be provided to all shelters, mess rooms, sanitary conveniences and other facilities.

Storage of materials

Equipment must not be allowed in mess rooms and shelters.

Lead compounds

When operatives use lead or lead compounds on site the washing facilities must be improved to the scale of one wash basin, bucket, etc., for every 5 men and nail brushes must be provided.

Sharing facilities

On sites where more than one employer is working, or where there are several sub-contractors, each employer technically must provide his own employees with the relevant facilities.

It is obviously sensible, when possible, to avoid a shanty town of individual mess rooms, toilets, etc., and accordingly the regulations make provision for the sharing of facilities by employees of different contractors. The following can be shared:

1. First-aid boxes; first-aid rooms and qualified first-aiders.
2. Ambulances, stretchers.
3. Shelters, messrooms, canteens.
4. Washing facilities.
5. Toilet facilities.

If facilities are to be shared then the employer who *provides* the facilities must:

1. Ensure that the facilities meet the requirements for the total who will use them.
2. Keep a register, Form 2202 Part B, showing the name of the firms sharing his facilities and what facilities are shared.
3. Issue a certificate, Form 2202 Part A, to each firm sharing his facilities and detailing which facilities are, in fact, being used.

Health hazards

Reference has been mde in previous chapters to the dangers associated with dust and fumes, particularly welding fumes; with lasers and with the hazards of radiation. These are all very real hazards but mostly brought on by the use of somewhat specialist equipment.

Industrial dermatitis

Dermatitis, on the other hand, or on either hand come to that, is far easier to contract and is a major source of sickness in the construction industry, some estimates suggesting that in excess of

200,000 working days are lost every year. It is neither infectious nor contagious but can be chronic and very distressing to the sufferer.

Skin care

Skin is the largest single organ of the body and, far from being a mere covering for our flesh and bones, is a living part of us. The causes of skin trouble in the construction industry usually fall within the following list but, as there is so much variation in different people's reactions to substances, it is very difficult to diagnose particular cases:

- cement
- paints, varnishes, etc.
- certain woods, mainly hardwoods
- epoxy and acrylic resins
- tar, bitumen
- solvents of all kinds
- acids and alkalis
- brick and plaster dust.

The basic need is to prevent contact or, where this is impracticable, to reduce the period of contact to the minimum. Personnel employed on any process where there is a possible skin hazard should be told about it and instructed in the right technique to minimize the risk.

Good housekeeping is essential. A person working in slovenly surroundings can be expected to be careless in his approach to personal hygiene and it is a lack of personal hygiene that so often leads to skin diseases. Gloves and aprons help to reduce contact but must be kept clean, as the finest protective equipment available can be useless if allowed to become contaminated.

Persons exposed to known skin irritants and dermatitis hazards should be provided with barrier creams but care must be exercised to ensure that the correct barrier for the risk is employed.

An efficient and properly formulated skin cleanser must be provided and the use, or misuse is the better word, of solvents, detergents and bleaches prohibited for washing the skin.

Finally the supervisor should institute a planned and regular examination of work methods, wherever there is a risk of skin disease and should be on the lookout for evidence of dermatitis. Any signs of the disease, such as reddening or blistering of the skin, should be treated seriously as early medical attention is essential.

Vibration induced white fingers

The operators of portable vibratory tools can be exposed to a condition known as vibration induced white fingers or VWF for short. By no means everybody is susceptible but it is important that operators are given information about the condition as there is some evidence that it can become progressive and may spread from the fingers to affect other parts of the circulation.

The first sign is that the tips of the fingers go white and feel numb when the hands are cold. The fingers may also swell. If an operative who is susceptible continues to be exposed to vibration the condition can worsen to such an extent that any exposure to cold makes the fingers go numb. This can obviously be an occupational handicap and any operative so affected should be given other work.

25

Personal protection

In the perfect working environment the provision of personal protective equipment would be unnecessary, as all hazards would be eradicated at source. However, this happy situation is not yet with us and accordingly items of personal protection play an essential part in preventing injury accidents on site.

Legal position

Employers are obliged by law to provide the following items:

1. Eye protectors or shields – for specified operations.
2. Respirators – when dangerous fumes and dusts are present and adequate ventilation is not practicable.
3. Protective clothing for men working with asbestos products when adequate exhaust ventilation is not practicable.
4. Ear protectors – when woodworking machinery noise levels cannot be satisfactorily reduced and in other noisy conditions.
5. Normal wet weather protective gear when men are required to continue working in rain, snow, sleet or hail.

Working rules

In addition to the above items, which are all supported by regulations, the wearing of safety helmets is also strongly recommended by the working rules agreed by the employers' federations and the unions concerned.

The rules say:

1. It is the intention that the habitual wearing of safety helmets shall extend to all workplaces where a risk of head injury is present.

2. Consequently, except where company or site rules or notices make exception, or where the foreseeable risk of head injury is negligible, operatives are required to wear safety helmets at all times when in the vicinity of construction operations.
3. Subject to the use of appropriate disciplinary procedures under the working rule, action taken by an employer to enforce these provisions shall be supported by all the parties to the agreement.
4. It is open to employers and operatives on any job to agree on the arrangements for the provision and maintenance of safety helmets.

It will be appreciated that members of supervisory staff will not normally be embraced by these working rules but will generally be required to comply with the suggested standards by their terms of employment. It is essential that a good example is set by staff in the wearing of safety helmets and indeed in the wearing of any required protective clothing or equipment.

Good practice

Whilst the wearing of safety footwear is neither required by regulation nor by working rules, it is obviously desirable that footwear is suitable.

Some companies institute arrangements whereby safety footwear can be purchased by employees at subsidized prices, or at least at wholesale prices. If your company does not provide this service possibly you could suggest that suitable arrangements should be made.

The provision of gloves also falls into the grey area between legal requirements and good practice but it is obviously sensible, and good practical economics if suitable gloves are provided when the handling of sharp, rough or abrasive materials is necessitated. New steel reinforcement rods and sheet piles are notorious for sharp edges and deserve special attention.

Returning to the legal aspects: the requirements for respirators when working with dangerous fumes and dusts; the need for protective clothing in some instances of asbestos work; the provision of inclement weather clothing; all these points have been discussed in previous chapters.

We now deal with the remaining major items, namely eye protectors, ear protectors and hearing conservation.

Eye protection

The Protection of Eyes Regulations 1974 is a short code of regulations, of only eleven sections, followed by some very detailed schedules which list the various operations for which eye protection is required.

The regulations require that suitable eye protection or shields are provided for the protection of persons employed in processes listed in the schedules. The protection provided must conform to approved standards and be issued on a personal basis except in the case of persons only occasionally employed. There is also a requirement to provide independently mounted shields for the protection of persons not directly involved in a process but required to work in a position where their eyes are at risk from particles thrown off or from the intense light produced by the process.

The major items from the schedules to the regulations are listed below, particularly those that are likely to concern construction work.

Processes in which approved eye protectors are required

1. Shot blasting of concrete.
2. The cleaning of buildings or structures by means of shot or other abrasive materials propelled by compressed air.
3. Cleaning by means of high-pressure water jets.
4. The striking of masonry nails by means of a hammer or other hand tool or by means of a power-driven portable tool.
5. Any work carried out with a hand-held cartridge operated tool.
6. The chipping of metal, and the chipping, knocking out, cutting out or cutting off of cold rivets, or similar articles from any structure or plant, by means of a hammer, chisel, or similar hand tool, or by means of a power-driven portable tool.
7. The chipping or scurfing of paint, scale or rust by means of a hand tool or by means of a power-driven portable tool.
8. The use of a high-speed metal cutting saw or an abrasive cutting-off wheel or disc, which in either case is power-driven.
9. The dismantling or demolition of plant which contains or has contained acids, alkalis or dangerous corrosive substances.

10. Injection by pressure of liquids or solutions into buildings or structures.
11. The breaking, cutting, cutting into, dressing, carving or drilling by means of a power-driven portable tool or by means of a hammer, chisel, pick or similar hand tool other than a trowel, of any of the following, that is to say:

    (a) glass, hard plastics, concrete, fired clay, plaster, slag or stone (whether natural or artificial);
    (b) materials similar to any of the foregoing;
    (c) articles consisting wholly or partly of any of the foregoing;
    (d) stonework, brickwork or blockwork;
    (e) bricks, tiles or blocks (except blocks made of wood).

    It will be noted that a bricklayer may cut bricks with his trowel but, strictly speaking, may not do so with a hammer and chisel unless he is properly protected. The supervisor may well find some difficulty in enforcing this particular section but should do his best to ensure compliance by providing suitable goggles and by pointing out the possible dangers to those concerned.

Processes in which approved shields or approved fixed shields are required

12. Any process involving the use of an exposed electric arc.

Processes in which approved eye protectors or approved shields or approved fixed shields are required

13. The welding of metals by means of apparatus to which oxygen or any flammable gas or vapour is supplied under pressure.
14. Any process involving the use of an instrument which produces light amplification by the stimulated emission of radiation (laser), being a process in which there is a reasonably foreseeable risk of injury to the eyes of any person engaged in the process from radiation.
15. Truing or dressing of an abrasive wheel.

Cases where protection is required for persons at risk from, but not employed in, a specified process

16. The chipping of metal and the chipping, knocking out, cutting out or cutting off of cold rivets, or similar articles from any structure or plant by means of a hammer, chisel,

or similar hand tool, or by means of a power-driven portable tool.

17. Any process involving the use of an exposed electric arc.
18. Any process involving the use of an instrument which produces light amplification by the stimulated emission of radiation (laser), where in any such process there is a reasonably foreseeable risk of injury to the eyes of any person not engaged in the process from radiation.

Employed persons' responsibilities

It will be appreciated, from the lists of operations quoted, that the legal requirement to provide goggles is very extensive and that the provision of suitable and acceptable protection is very important. The employed person is required by the regulations to wear the equipment that is provided and to take reasonable care of it but, even so, if uncomfortable equipment is provided it will *not* be worn and will end up being thrown away, leaving unprotected eyes and adding further costs to the employer.

Noise and hearing conservation

Noise has been described as unwanted sound. In addition to causing annoyance, noise may interfere with working efficiency by inducing stress and by disturbing concentration, especially where the work is difficult or highly skilled. Additionally noise may hinder communications and mask warning signals and calls and so could be the cause of accidents.

Probably more important, it may damage the hearing of employees. A short exposure to high sound levels may result in a temporary hearing loss, lasting from a few seconds to a few days, while regular exposure to far lower sound levels over a long period of time may result in destruction of certain inner ear structures and a consequent loss of hearing which is permanent and incurable.

The long-serving employee who believes he has 'got used to the noise' has almost certainly sustained a hearing loss.

Acceptable noise levels

Noise is measured in decibels and can only be established by using a noise level meter. Exposure to high noise levels, whilst not in any way desirable, can usually be undertaken for short periods without permanent damage.

The table below gives a guide to the recommended maximum exposure without wearing ear protection:

Average sound pressure level dB(A)	Maximum exposure hours
90	8
93	4
96	2
99	1
102	½
102 +	*nil*

What should be appreciated is that an increase of 3 dB(A) indicates a doubling of the sound power, hence the necessary halving of recommended exposure hours shown in the table.

Actual sound pressure levels for particular items of plant and operations vary tremendously, depending on site conditions, but it is quite usual for a level in excess of 90 dB(A) to be found when working close to heavy lorries or plant, particularly pneumatic drilling equipment and piling work or when in confined areas.

Methods of controlling noise exposure

Measures to reduce workers' exposure to noise should start with quietening plant and tools as far as is practicable.

Exhaust muffles and sound-reducing covers for such items as compressors and pumps can be helpful; a wall of straw bales can be very effective; good maintenance of vehicles and machinery can also cut down much unnecessary noise.

Particular care should be taken to protect those workers who are exposed to noise for the whole day or shift. When the time of exposure can *not* be reduced suitable ear defenders may be the only remedy, but it must be emphasized that these are the last line of defence and should only be considered after all else has been considered.

It should also be appreciated that the standards referred to previously are *maximum* limits and that some individuals, who are particularly sensitive to noise, may find the permitted level intolerable. Complaints from individuals about noise should therefore be taken seriously and not disposed of by the expedient of referring to a sound level meter. Employees who are so affected should if possible be transferred to less-noisy work situations, or supplied with ear defenders even though the noise level may not make this necessary for their colleagues.

26

Manual handling

A comic song, popular some years ago, described a chap called Fred and his pals handling an undefined but obviously heavy piece of furniture. Fred's impetuosity led to his receiving, in the words of the song, 'half a ton of rubble on the top of his dome'. This we assume caused, at the very least, a nasty scalp wound and would probably be listed by the statistician as a 'handling' accident.

Official accident figures reveal that 'handling' causes many injuries in the construction industry and it is certainly regularly listed as one of the 'big five' causes. The sort of injuries that arise are these; hernia, torn back muscles, slipped disc and various cuts, bruises and crush injuries to the hands, arms, feet and legs.

Yet manual handling is the operation above all others that is most under the control of the operative. How then are these accidents caused? Almost all arise from the failure of the operative to:

1. Appreciate the mechanics of the human body.
2. Understand the nature of the load to be handled.
3. Put into practice correct handling techniques.
4. Make use of protective equipment.
5. Use necessary care and foresight.

It falls on the site supervisor to ensure that the operatives on his site do not have these failings and that training is provided in all aspects of manual handling.

Simple body mechanics

In the lifting context the human body can be compared with a crane. A crane is far more stable and able to handle heavier loads when the load is close at hand and the jib can be used in an almost vertical position. A similar situation exists with the human frame, much greater lifting power being available when a load is held tight

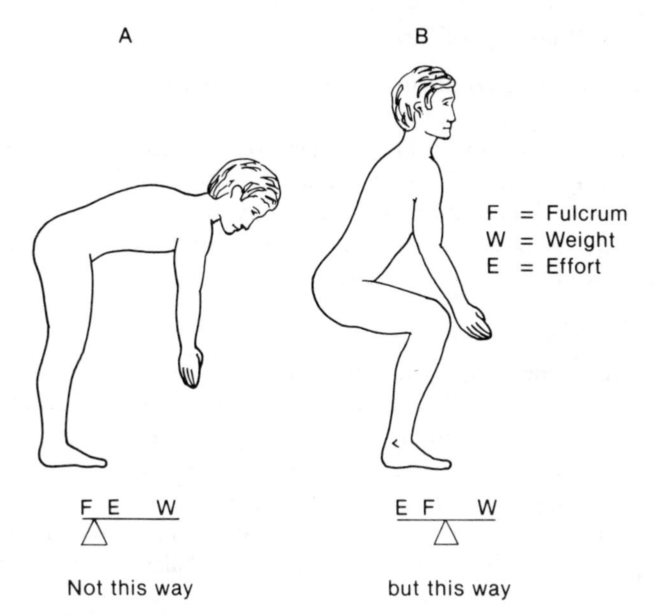

Fig. 26.1 Lifting mechanics. Keeping the weight and the fulcrum close together as in B gives a tremendous mechanical advantage.

to the body and the back is kept straight.

It must be appreciated that when a person lifts incorrectly by bending over from the waist he must straighten his back during the lifting operation and so has to lift, not only the load, but also the considerable weight of the top half of the body (see Fig. 26.1).

General safety rules

1. The supervisor should ensure that an operative never tackles a load that is beyond his capacity. If there is any doubt, always use two men.
2. Full use should be made of mechanical aids.
3. Extra care must be taken with awkwardly shaped objects; the weight which a person can safely lift decreases as the awkwardness of the shape increases.
4. When lifting is to be done by more than one man be sure to appoint a team leader to give instructions.
5. Ensure that all necessary protective equipment is available and used.
6. Consider the underfoot conditions – slippery sites lead to many handling accidents.

Principles of lifting

Having considered the general safety rules it becomes necessary to look at the specifics of safe lifting.

The winning slogan in a competition organized some years ago ran 'Back straight, bended knees, lift with safety and with ease'. Besides 'back and knees' there are, however, other points to consider and in fact it is generally accepted that there are six aspects which lead to safe and effective lifting.

Correct position of feet

A lifting operation is usually a preliminary to some degree of movement with the load and so it is important that the leading foot should be placed pointing in the direction of travel with the other foot ready to move. The actual distance between the feet will, of course, vary depending on the height of the operative but an average distance is about 500 mm. Positioning the feet in this way helps to avoid those strains caused by overbalancing and twisting.

Back straight

If the back is kept straight, the pressure imposed on the lumbar intervertebral discs is evenly distributed, but if the back is bent it is inevitable that certain sections of the discs are strained. Keeping the back straight does not necessarily mean it must be vertical; but by bending the knees and using the large thigh muscles to do the work of lifting, the back can be kept in its natural 'straight' position (see Fig. 26.2).

Correct hold

A good grip, using the palms of the hands and not just the finger tips, is essential. Make use of gloves whenever necessary and take care not to pinch fingers when putting down the load.

Position of arm

The arms should be kept as straight as possible to avoid excessive strain on the muscles of the upper arms and the load being carried or lifted should be kept as close as possible to the body.

Position of head

It may be thought that the position of the head is unimportant in lifting technique but the action of keeping the chin tucked in helps to keep the spine straight and so is important.

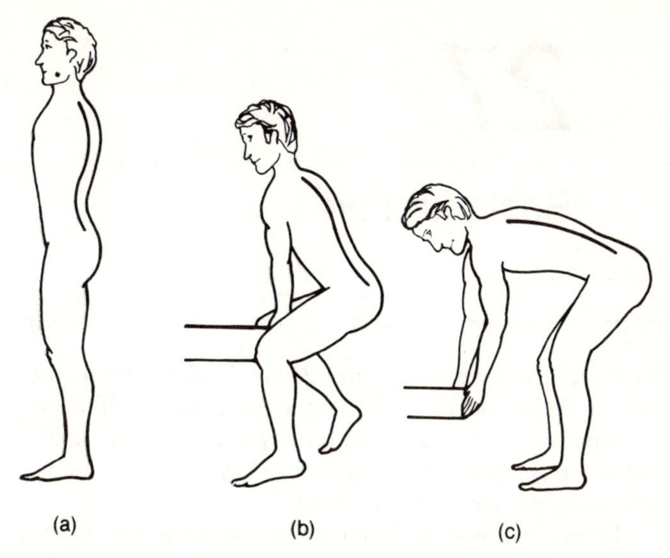

Fig. 26.2 Shape of spine. (a) The natural shape of the spine when standing. (b) the natural shape being maintained when lifting correctly. (c) The bending of the spine when lifting incorrectly.

Use of body weight

The body weight can be used in many lifting operations for, as it were, knocking the load off balance and putting it into an easier position for lifting.

An awkward heavy crate, for example, needs to be rocked over onto an edge to enable a grip to be obtained and an oil drum must on occasion be pulled over to a balance point to facilitate rolling.

The important thing in all uses of body weight is to ensure that balance is maintained and this is best achieved, as already mentioned, by careful positioning of the feet.

The universal practice of these simple rules would drastically reduce the number of handling accidents in the industry and would lead to greater efficiency on contract sites. A lifting operation carefully planned is completed quickly and not left half finished because of the interruption caused by an accident.

To revert to our comic song: Fred and his pals never did finish their job – they left it on the landing!

27

Basic first aid

We have discussed in Chapter 24 the legal requirement to have a qualified 'first-aider' on the larger sites, but it will be appreciated that injuries can be sustained on any site however small it may be and accordingly I believe that we all have a duty to understand the basic facts of elementary first aid.

The most important point to remember is that *breathing failure* and *bleeding* require immediate attention. Everything else can, and should, wait.

Breathing

If oxygen is cut off from the brain for longer than about four minutes the brain will probably become permanently damaged and if this interruption is for *much* longer then death will occur. The oxygen needed for the brain is carried by the blood which is pumped to the brain by the action of the heart, having picked up the necessary oxygen from the air supply inhaled by the lungs, and so it will be seen that the oxygen shortage can be brought about by three main causes:

1. A failure of the heart in its pumping action.
2. Failure of the lungs.
3. An obstruction of some sort in the air passages, i.e. mouth, nose and windpipe, preventing air from being drawn into the lungs.

An obstruction in the air passages

In the normal everyday situation it is probably the last of the three suggested causes that is most common and even on construction sites there is always the possibility of some unfortunate person choking on his ham sandwich. There are also other situations to which the construction worker is vulnerable. The soil in trench excavations, which collapse and trap an operative, can easily exert sufficient force on the chest and stomach to prevent breathing, and similar 'trapping' accidents can have the same consequences.

Speed is obviously essential in all such cases. Accidents have been reported when men were trapped to just above the waist by a fall of earth but were not rescued with sufficient speed and died of suffocation.

Other accidents can result in the far more dangerous situation of the airway being obstructed in a person who is unconscious. It is quite common for persons who fall from heights, or who are struck by plant or equipment, to be knocked unconscious.

If they then are left lying on their back, whilst in this state of unconsciousness, their tongue can easily drop back, block the windpipe and stop them from breathing. False teeth and vomit can produce the same problem.

It is essential, therefore, when faced with this situation, that you quickly remove any sickness and false teeth from the mouth and make sure that the tongue is not lying across the windpipe. Then the patient should be placed on his front with his face turned to the side. This is called the recovery position (see Fig 27.1).

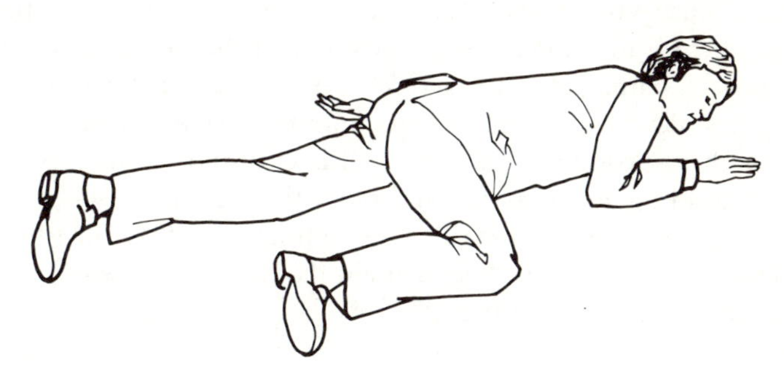

Fig. 27.1 The recovery position.

The recovery position

Ensure that the tongue will fall naturally away from the windpipe and that any sickness will not block the air passages. No attempt should be made to move a patient into the recovery position, however, if it appears likely that he has a fracture of the spine or neck. Any bending of the spine or neck may cause severe damage to the spinal cord and subsequent permanent paralysis so extreme care must be taken in dealing with such a patient.

If, however, breathing does not restart when the obstructions are removed, or the ham sandwich is coughed up, it will be necessary to start artificial respiration to restart the lungs working.

Artificial respiration

There are several methods of artificial respiration but the best known is the 'kiss of life' or mouth-to-mouth method. This is very simple and can be learned and performed by children as well as adults.

First turn the patient onto his back and lift his neck and put a folded coat or similar article under his shoulders. This enables you to lift his head well back to open the air passages. Make sure there is no obstruction in the mouth and that the tongue is not blocking the wind-pipe and then pinch the patient's nostrils closed, place your mouth over his and blow in until you see his chest rise. Remove your mouth and you should hear an outflow of air and see the chest lower. Repeat the process naturally as the lungs empty, which will usually be at the rate of about 10–12 times a minute for an adult and about 20 times a minute for a child. Continue until your patient starts to breathe for himself and then roll him gently into the recovery position. If the mouth of the patient has been damaged in the accident, or you are unable to get an airtight seal when you place your mouth over his, you can use the mouth-to-nose method. The drill is exactly the same except that you hold his lips together and blow through his nose.

Please remember that this is not a time to be fussy or squeamish. You have no time to lose if brain damage, and possibly death, is to be prevented. (see Fig. 27.2).

Failure of the heart

When the heart stops beating no blood is pumped to the brain and consequently the oxygen necessary to keep the brain alive is cut off.

Apart from the 'heart attack', or coronary thrombosis as it is known medically, the heart can be stopped by other causes and the most common of these that would happen on a construction site are:

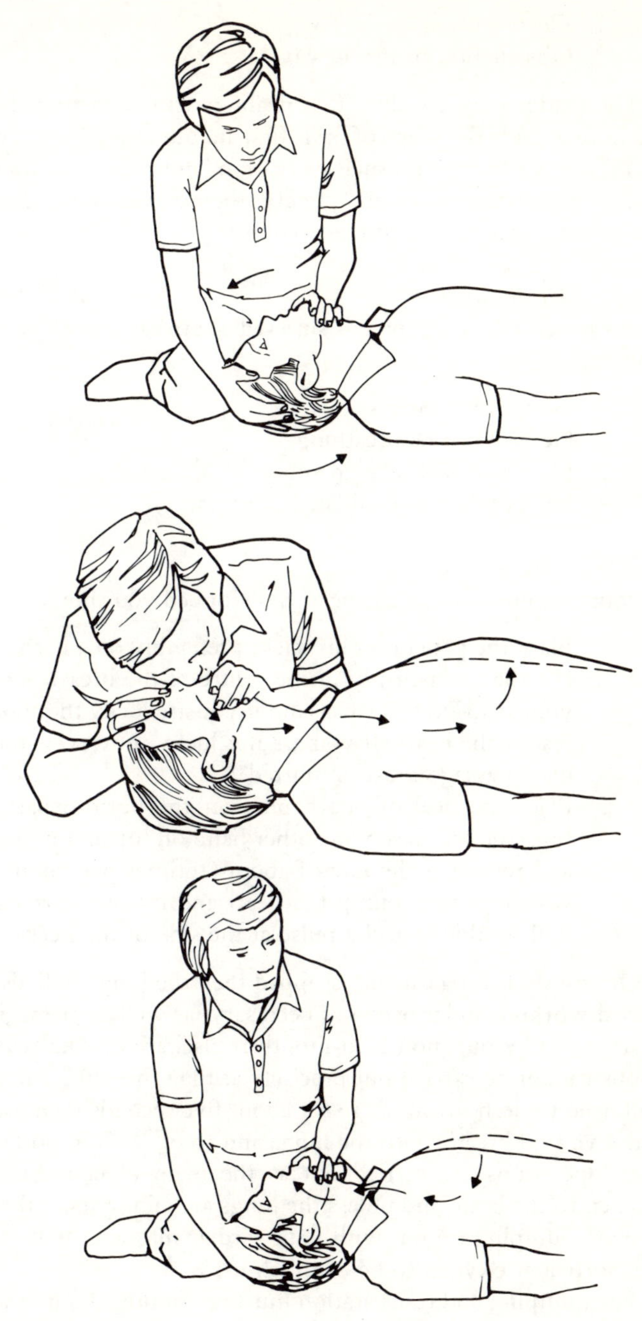

Fig. 27.2 Kiss of life.

1. Electrocution.
2. Obstruction of the airway.

The reader may consider that dealing with restarting the heart is a little outside the scope of basic first aid but everything is relative and if you are faced with such a situation, and there is no likelihood of expert assistance arriving quickly, then you have no alternative but to offer whatever assistance you can.

The diagnosis

You can tell that a person has suffered a 'cardiac arrest' by the following:

1. He will be unconscious.
2. He will not be breathing.
3. No pulse will be evident.
4. His pupils will be dilated (very large).

The treament

If your diagnosis is that the heart has stopped you should:

1. Place the patient on his back, preferably on a hard surface.
2. Hit the chest firmly in the centre of the breast bone with your fist. Do this twice and if you are lucky the shock will restart the heart. If you are not lucky, however, then cardiac massage must be started.
3. Place the heel of one hand over the bottom part of the breastbone, place your other hand on top and press firmly and release at the rate of about 60 times per minute. This will, in effect, pump the heart and if done correctly you will be able to feel a pulse at the side of the neck.

Obviously if the heart has stopped then the lungs will also have ceased working and it becomes necessary, as well as massaging the heart, to carry out mouth-to-mouth resuscitation. The two operations cannot be carried out precisely at the same time and so you will need to do heart massage for about five seconds then break off and give two breaths into the lungs and so on. It is important that both operations are carried out as the main objective is to get oxygen to the brain and unless the lungs are taking in air the blood you are pumping round, with your cardiac massage, will not contain sufficient oxygen to be of much use.

The pumping and resuscitation must be continued until the heart and lungs operate for themselves or until a doctor arrives and takes over your patient.

Bleeding

The control of severe bleeding is the next important action that must be taken and again speed is essential. Lay the patient down and apply pressure to the wound. Exactly how this pressure is applied depends on the type and size of the wound. It may be possible to control the bleeding by squeezing the sides of the wound together and this is often the first step to be taken. Next a pad should be applied over the wound and bandaged on firmly. If the first-aid box is handy then obviously the clean pads and bandages from there should be used but if this is not the case then use the cleanest thing you can find. Never attempt to remove anything which is firmly fixed in the wound as this will possibly increase the bleeding. If the pad and bandage do not stop the bleeding bandage another pad on top. Do *not* remove the original bandaging. Some reduction in bleeding can be achieved by raising the affected part, if it happens to be a limb, but on no account adopt this measure if there is any likelihood that the limb is also fractured.

Fractures

Fractures are quite common on construction sites and the golden rule to remember is to avoid aggravating the injury by moving the patient more than absolutely necessary.

Spine or neck

If there is any possibility of a fracture or injury to the spine or neck DO NOT MOVE your patient at all, provided he is breathing and is not at risk where he is lying. 'Do not move' includes sitting him up, which you should never do. There are tragically many permanently paralysed people who might have escaped such serious injury had not some well-meaning but ill-advised rescuer helped them to 'sit up'.

If he is at risk where he is then avoid bending his neck or back at all costs when you move him.

Fractured skull

If you suspect a fracture of the skull keep the patient quiet, lying down and await the ambulance. If he is unconscious turn his head to one side to allow any secretion to drain from his mouth.

Fractured limb

Fractures to the limbs are less serious than those previously mentioned but can also be aggravated easily by carelessness. It is often not at all obvious that a fracture *has* occurred but if you have any suspicion then you should assume the worst and treat accordingly.

A fracture of the arm is best treated by supporting the arm in a sling and transporting the patient to the local casualty department for treatment.

Fractures to the leg basically fall into two categories – ankle injuries and others. In the case of a fractured ankle it is usually possible for the patient to hop or be carried to some transport, but other leg fracture cases should not be moved until the limb has been immobilized. If only one leg is affected you should tie the patient's ankles together, using the good leg as a splint, and await the ambulance.

Electrocution

A severe electric shock may cause the lungs to stop breathing and the treatment in these cases is to give mouth-to-mouth resuscitation.

It may seem rather an obvious point to make but do make sure that the electricity supply has been switched off or that the patient has been dragged clear of the danger before touching him. There may be burns to the body but remember that these are of secondary importance to the task of starting the lungs working again.

Burns

The treatment for burns and scalds is immediately to immerse the damaged part of the body in cold water. Prompt action of this sort will certainly ease the pain and may also reduce the severity of the burn. Severe burns should then be covered with a clean cloth, a towel or something similar, and the patient taken to hospital.

If you happen to be present when, for some reason, a person's clothes catch fire, immediately try to lay him down and smother the flames with a coat or something similar of a non-flammable nature. The act of laying him down reduces the area of his body that the rising flames can reach.

Remember

The maintenance of breathing and the control of serious bleeding are your first priorities.

If oxygen is prevented from reaching the brain for more than about four minutes serious brain damage will almost certainly result.

Make sure that you know and remember what to do to restart the lungs and the heart. You never know when you may be called upon to demonstrate this skill.

28

Fire prevention and control

Fire is not one of the major causes of injury accidents on construction sites but this is largely because an occupied building has much more to offer in the way of fuel for a fire than the same building has whilst under construction.

Whilst it is true to say that flammable materials are used in the construction process it is also the case that a fully furnished office block or factory building will have many more risk materials, additional risk areas and, probably most important, more people. It is, after all, in most cases people who start fires.

However, there are regularly small fires on construction sites, which occasionally become large fires and make the headlines, and it is the duty of the site supervisor to understand the basic principles of fire prevention and control so as better to protect his workforce and his project from these dangers.

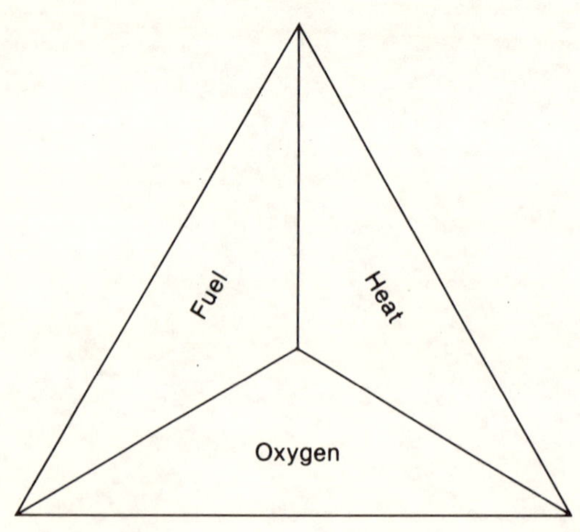

Fig. 28.1 The triangle of fire. Fire has three essential ingredients: fuel, oxygen and heat.

Understanding fire chemistry

The site supervisor's approach to fire prevention and control should begin with an understanding of basic fire chemistry and the first point to understand is that there are three elements necessary to start a fire – fuel, oxygen and heat. These are perhaps best remembered by reference to the 'triangle of fire' (see Fig. 28.1).

The removal of one or more sides from the triangle causes the fire to go out (see Fig. 28.2).

This can be done by:

1. Cooling the fire to remove the heat, usually with water.

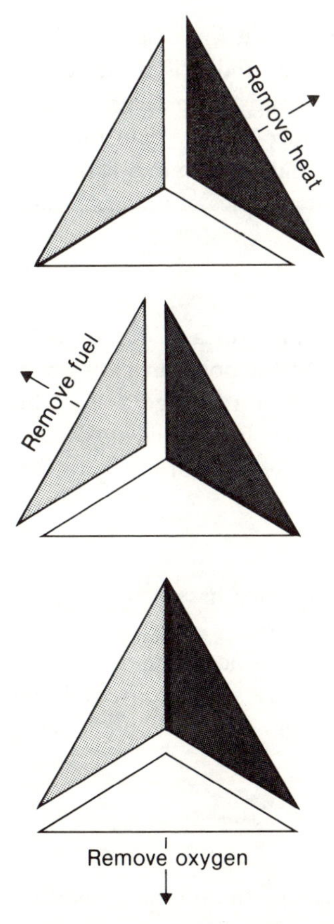

Fig. 28.2 Breaking the triangle. If one or more elements is removed, the fire will be extinguished.

2. Starving the fire of fuel.
3. Smothering the fire by limiting its oxygen supply.

One way of smothering a fire is to drive away the air from near the fire and replace it with carbon dioxide or other gases which do not support combustion. This is what happens when a carbon dioxide extinguisher is used. Another method is to use a dry chemical powder, or a vaporizing liquid extinguisher and these operate by interfering with the chemical reactions of the flame.

How fuels behave in fire

Solids

Before a combustible solid will burn it must be heated sufficiently for it to give off vapours and gases and it is these which burn in the characteristic flames. Those solids with large surface areas in relation to their volume burn more readily than those which are more compact. Thin sheet material, corrugated cardboard, wood shavings, these are all common examples of materials having large surface areas.

Dusts

Combustible dusts have a *very* high surface area to volume ratio and are particularly hazardous. Dust deposits in a joinery workshop can spread fire across a room or along a ledge or roof beam very quickly or, on the other hand, can be ignited unknowingly and smoulder slowly for long periods before springing to life.

Liquids

As with combustible solids, a vapour has to be produced at the surface of a liquid before it will burn.

Some flammable liquids need to be heated in order for sufficient vapour to be produced; examples are white spirit and fuel oil, but others, such as petroleum spirit, give off ignitable vapours at well below normal temperatures.

Fire causes

Let us look at the common causes of fire on our construction sites. These are not listed in any particular order.

Smoking and matches

The careless disposal of cigarette ends and matches is a regular cause of fires. Insistence on special smoking areas is difficult on site but this type of control is very important when highly flammable substances are being used.

Faulty or misused electrical equipment

Overloading of circuits and sockets, damaged wiring and equipment, improvised connections, all can lead to a build-up of heat and sparking sufficient to provide a source of ignition.

Welding activities

The containment of sparks is very difficult, particularly on multi-level construction contracts, and all types of welding and cutting operations must be carefully monitored.

LPG fired equipment

The particular hazards of this type of equipment have already been covered in a previous chapter but the fire potential needs to be repeated here. The proper supervision of the installing and use of equipment is essential if gas leaks and subsequent fires are to be avoided.

Combustion sparks

Rubbish burning must also be properly controlled, as must unauthorized warming braziers that often tend to appear on sites in extreme weather conditions. In very cold weather the sensible supervisor will make provision for additional breaks for his men to enable them to warm themselves. This is much to be preferred to allowing the setting up of 'roast chestnut' stalls around the site.

Legal requirements

On contract sites where more than twenty people are employed in an office complex, or more than ten on any other than the ground floor, you will need to obtain a 'safe means of escape certificate' from the local fire authority. This will also entail providing a fire-alarm warning system, adequate fire extinguishers and training in the use of the extinguishers for the employees.

The requirement to train is reinforced by Section 2.2(c) of the Health and Safety at Work Act which says that the employer must provide such information, instruction, training and supervision as is necessary to ensure, so far as is reasonably practicable, the health and safety at work of his employees.

This requirement to train turns up so regularly that I hope the reader will eventually begin to appreciate its importance, and in the context of fire fighting it is very important.

There have unfortunately been a number of accidents where untrained persons have put themselves at risk while attempting to deal with a fire and have sustained serious injuries as a result. No one should be told to attack a fire unless they have been given adequate training and, in fact, they should be discouraged from doing so. The supervisor who disregards this advice can find himself in legal difficulties if an untrained 'fire fighter' is injured whilst tackling the blaze.

Types of extinguishers and use

Training in the use of fire extinguishers should be given by experienced personnel. You may have company safety officers capable of undertaking this duty or you may have to import some assistance from either the local fire authority or one of the suppliers. The notes that follow are basic information and *not* sufficiently detailed to be used as a training manual.

General notes

Most outbreaks of fire have small beginnings but can spread very quickly, so a speedy attack is essential.

Only attempt to fight a fire if you have been trained to do so and never continue to fight a fire if:

1. It is dangerous to do so.
2. There is a possibility that your escape route may be cut off by fire or smoke.
3. The fire continues to grow in spite of your efforts.
4. There are liquefied petroleum gas cylinders threatened by the fire.

If possible never attempt to fight a fire alone. There may obviously be instances when a lone attempt is unavoidable but in these cases even more consideration must be given to the points 1 to 4 above.

If you have to withdraw from a fire within a building, close windows and doors behind you whenever possible.

Water extinguishers

Water-filled extinguishers are suitable for use on most fires except those involving flammable liquids or live electrical apparatus.

Take up a position where a quick and safe retreat is possible. Crouching low will help you to keep clear of smoke and allow a closer approach. Direct the water jet at the base of the flames. A fire spreading vertically should be attacked at its lowest point and followed up.

Always ensure that a fire is completely extinguished and not still smouldering and liable to reignite.

Foam extinguishers

The purpose of using a foam extinguisher is to put a barrier of foam between the surface of the fire and the air and so extinguish the flame. They are suitable for most fires involving flammable liquids. They can also be used on ordinary fires of solid materials but will not be as effective as water. They should not be used on fires where live electrical equipment is involved. When used on flammable liquid fires they are only really effective if the liquid is in a container of some sort and are not suitable if the burning liquid is running.

The method of using a foam extinguisher is either to aim the jet at a vertical inside edge of the container above the level of the burning liquid, which will break up the jet and allow the foam to spread across the surface of the liquid or, if this is not possible, to direct the jet with a gentle sweeping motion to allow the foam to drop down and lie on the burning surface. Advancing too close and aiming the jet directly into the burning liquid will be ineffective and may also cause the burning liquid to splash from the container onto the surroundings.

Carbon dioxide, dry powder and halon extinguishers

These types of extinguisher are designed for dealing with fires involving flammable liquids or electrical apparatus. They are, however, safe for use on all fires, although they will not prove to be as effective as those types specially designed, and are particularly useful when dealing with mixed fires.

The method of operation when dealing with flammable liquid fires is to direct the horn towards the nearest side of the fire and, with a sweeping movement, drive the flame away towards the far edge until completely extinguished.

When dealing with fires involving electrical equipment it is always preferable to switch off the electricity supply whenever possible and then you should direct the horn of the extinguisher straight at the fire.

Fire prevention planning

It is at the planning stage of the contract that most can be achieved in respect of fire prevention.

The most regularly involved locations for site fires are:

1. Site offices, messrooms, drying rooms.
2. Storage huts and external storage areas.
3. Work areas where welding operations are carried out.
4. Areas where rubbish is burned.

Dealing with these in order:

Site Buildings

Be sure that electrical installations are carried out by a competent electrician and that any liquefied petroleum gas apparatus is properly fitted and tested for leaks before its being put into service. Where the floors of site huts are raised above ground level the space beneath them should be enclosed to prevent an accumulation of rubbish. Whilst the idea of linking individual huts together is attractive from a convenience point of view it is to be opposed from the fire-prevention angle and huts should be adequately spaced to lessen the danger of fire spreading.

Particular care must be taken in drying rooms to ensure that clothing cannot be hung too close, or draped over, heaters.

Storage areas

Storage of flammable materials should be planned to be kept to the minimum possible. Due consideration must be given to the separation of stores areas and the special storage requirements for highly flammable liquids and liquefied petroleum gases.

Provision should also be made for easy access for any necessary fire fighting equipment.

Welding operations

Welding operations and blow lamps constitute a direct source of ignition. The site supervisor should ensure that the immediate work area is cleared of all combustible materials and, if it is possible, provide asbestos or metal screens to contain any sparks or hot debris.

Burning rubbish

All sorts of rubbish can and does accumulate on a construction site and a good deal of it makes first rate fuel. Wood offcuts and shavings; packing materials, paint scrapings, floor sweepings and so on all need regular attention or they will build up and just ask to be lit. A proper programme of waste and rubbish clearance is an essential part of any fire prevention planning and should be high on your order of priorities.

A final word. Fire fighting is a professional occupation and, apart from the small conflagration, best left to the fire brigade. Fire prevention is, however, quite different and something in which we can all play a part. Let us all 'know our place', concentrate on creating a fire-safe environment and consequently avoid those dangers associated with the fighting of fires.

29

Protecting the public

Section 3 of the Health and Safety at Work Act requires every employer to conduct his undertaking in such a way as to ensure, so far as is reasonably practicable, that persons *not* in his employment but who may be affected thereby are not exposed to risks to their health and safety.

Accordingly members of the public living near your site, passing by it or visiting it, all have a right to expect to be safe and not to be subjected to health hazards. This even applies in a limited degree to a person who is trespassing on site for, while the civil law accepts that he enters and remains at his own risk, Section 3 of the Health and Safety at Work Act makes no such exclusion for the trespasser. Happily it is not considered likely that criminal proceedings would be taken against an employer who fails in this duty unless he unwisely took to setting traps. This is not the case with children, however, who are expected by the law to take less care of themselves than would an adult. But more of that later.

We have in previous chapters discussed the need for adequate site security and also the special problems associated with demolition work, but one aspect of construction and demolition work that probably affects neighbours more than any other is that of noise.

Noise and the public

Noise from construction or demolition sites presents different problems of control compared with most other types of industrial operations since:

1. They are usually carried on in the open.
2. It is difficult to divorce the sites from areas that are particularly sensitive to noise.
3. Much of the noise is caused by machinery and there is a

tendency for machines in use to be larger, more powerful and consequently noisier than they were a few years ago.

However, if the noise created by your site is a cause of annoyance to people living and working nearby, you may find yourself forced into the noise abatement business whether you like it or not.

Control of Pollution Act 1974

This act deals primarily with waste disposal, water pollution and air pollution but does have two sections, Sections 60 and 61, aimed specifically at noise on construction sites. The act is controlled, not by the Health and Safety Executive but by the local authority and if they are approached by a person complaining about the noise emanating from your site, or they themselves think the noise level is unacceptable, you may find yourself served with a notice imposing requirements.

The notice may:

1. Specify the plant or machinery which is, or is not, to be used.
2. Specify the hours during which the work may be carried out.
3. Specify the level of noise which may be emitted from the site or from any particular part of the site.

These requirements could, as you will appreciate, make life very difficult for the contractor required to operate in a built-up residential area and the regulations, recognising this, make provision for the contractor to apply to the local authority to ascertain its noise requirements before work starts. The contractor does not have to do this but it is very sensible to do so in many cases as it reduces the risk of a local authority serving a notice after work has started with the consequential delay and additional expenditure.

If at the planning stage of a contract you think that complaints about noise may be forthcoming you should approach the environmental health department of the local authority at the earliest possible date; give them full detaiils about the work that has to be done, the methods you intend to adopt, the machinery that is likely to be used and, most important, the steps you propose taking to minimize the noise from your operations.

Provided the local authority is satisfied with your proposed steps to keep to an acceptable noise level it will give its consent.

Methods of controlling noise

There are two aspects of controlling noise that should be considered – controlling the source and controlling the spread.

Controlling noise at source

Much of the noise emanates from plant and machinery so care should be taken when siting plant. When machines are fitted with engine covers these should be kept closed as much as possible and full use should be made of any available mufflers or silencers.

In some extreme cases it may be necessary to enclose the machine within an acoustic shed (see Fig. 29.1). A shed can be extremely effective but, when machines are being silenced in this way, special provisions must be made for protecting any operator working within the acoustic shed.

Sheds such as the one illustrated are normally constructed from 6 mm plywood on timber framing and lined with about 50 mm of sound absorbent material such as glass fibre or woodwool slabs. Noise reductions of approximately 10 dB can be achieved in this way.

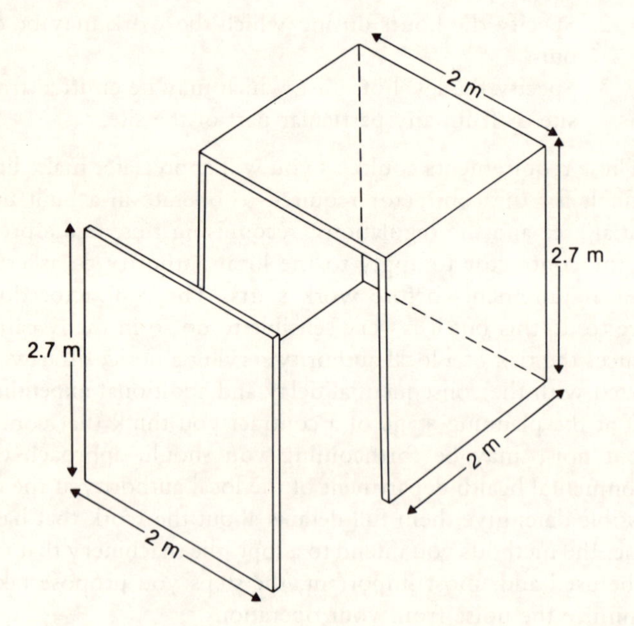

Fig. 29.1 Acoustic shed. Shed and screen constructed of 6 mm plywood, lined on the inside with 50 mm thickness of sound absorbent material.

Careful supervision plays a major part in controlling noise, thoughtless handling of materials, particularly when dismantling scaffolding, can create a lot of noise, dumpers and other items of plant left standing for hours, with their engines running, create unnecessary noise *and* waste money, and clear instructions should be given by the supervisor to combat items such as these.

Controlling the spread of noise

If noisy processes cannot be avoided, and controls at source prove to be ineffective or are impracticable, attempts should be made to limit the spread of noise either by introducing noise reduction screens or by making use of 'distance'.

Increasing the distance is often the most effective way of controlling noise but obviously has limitations. If a basement excavation is required at one end of a site the excavator needs to be there also – it cannot operate by remote control. There are some operations and items of plant, however, such as site workshops, sawbenches, etc., which do not usually require to be in any particular area and it may be possible, on occasion, to place these in positions where 'distance' can be used to advantage.

Noise reduction screens can be provided by making use of partially completed buildings, by positioning site offices, store buildings or temporary wooden screens between the noise source and the area required to be protected or by providing earth mounds or banks as a barrier. If an earth mound or a prefabricated screen is used, be sure to site the barrier as close as possible to the noise source, as it then becomes much more effective. Stationary items of plant, such as water pumps, can be effectively screened by siting them in a trench. This is particularly important if the pumps are required to work continously during the night.

Public relations

It is very important that good relations are maintained with people living and working nearby. If people are kept informed about noisy and dirty operations, particularly when explosives are to be used in demolition work, then much of the resistance and natural reaction to noisy construction operations can be overcome.

Protecting children

An information film, produced by the Health and Safety Executive, is entitled 'Building sites bite'.

The film, when released, received a mixed reception but, whether you like the film or not, there is no doubt that construction sites do have an attraction for children and that they can give nasty 'bites'.

We have a duty to protect all visitors to our sites but particularly children, even those who are technically trespassing. As has already been discussed, the civil law accepts that the unlawful guest must look after himself, but the Court of Appeal has also ruled that there is no general rule to be applied to all trespassers and each case depends on whether in the circumstances a duty was owed to the trespasser.

Obviously a greater duty is owed to children and, in the final analysis, the court will have to decide whether you, as occupier of the site, have done enough to save a trespassing child from harm.

Can we forget your legal liability for a moment and look at the moral situation. We all know that adventurous children will find a way of entering a site if enough determination is shown. The hoarding has not yet been designed that would keep them out and many sites in rural surroundings do not seem to rate a hoarding anyway. Site operations must be left as safe as we can possibly make them when work finishes for the day, particularly at weekends in the summer. Hoardings or fencing *should* be provided whenever possible and warning notices displayed but the precautions must not end there. Incomplete and unsafe scaffolding, unfenced excavations, badly stacked materials, plant not immobilized, accessible gas cylinders, unsafe electrical installations – none of these should be on your site in any case but do make doubly sure that they are not left for some trespassing, but none the less unsuspecting, child to find.

I hope you never have the job of knocking on a door and telling a wife that her husband has been killed while working for you on site. The author has had to do this in the past and will never forget it. There is only one task which might be harder to tackle and that is having to tell a mother that her child has been killed whilst *playing* on your site. You might be in the right, but you will always feel in the wrong.

Useful points of reference

British Standards Institution,
2 Park Street, London W1A 2BS.

HMSO,
49 High Holborn, London, WC1V 6HB.

Construction Safety manual published by National Federation of Building Trades Employers, 82 New Cavendish Street, London WIM 8AD.

Federation of Civil Engineering Contractors,
Cowdray House, 6 Portugal Street, London WC2A 2HH.

Construction Health and Safety Group,
John Ryder Training Centre, St. Ann's Road, Chertsey, Surrey, KT16 9AT.

Construction Industry Training Board,
Radnor House, London Road, Norbury, London SW16 4EL.

Royal Society for the Prevention of Accidents,
Cannon House, The Priory Queensway, Birmingham B4 6BS.

Health and Safety Executive
(See local telephone directory).

Index